STAR
FINDER
FOR BEGINNERS

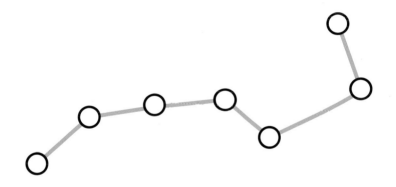

STAR FINDER

FOR BEGINNERS

A STEP-BY-STEP GUIDE TO THE NIGHT SKY

FOREWORD BY
DR. MAGGIE ADERIN-POCOCK

Editor Sarah MacLeod
Designer Louise Dick
Managing Editor Francesca Baines
Managing Art Editor Philip Letsu
Producer, Pre-Production Robert Dunn
Producer Gary Batchelor
Jacket Editor Claire Gell
Jacket Designers Surabhi Wadhwa, Juhi Sheth
Jackets Editorial Coordinator Priyanka Sharma
Managing Jackets Editor Saloni Singh
Jacket Design Development Manager Sophia MTT
Senior DTP Designer Harish Aggarwal
Picture Researcher Deepak Negi
Publisher Andrew Macintyre
Art Director Karen Self
Associate Publishing Director Liz Wheeler
Publishing Director Jonathan Metcalf

Contributor Ian Ridpath

First published in Great Britain in 2017 by
Dorling Kindersley Limited
DK, One Embassy Gardens, 8 Viaduct Gardens,
London, SW11 7BW

The authorised representative in the EEA is
Dorling Kindersley Verlag GmbH. Arnulfstr. 124,
80636 Munich, Germany

A CIP catalogue record for this book
is available from the British Library.
ISBN: 978-0-2412-8683-8

Printed and bound in China

www.dk.com

MIX
Paper | Supporting
responsible forestry
FSC™ C018179

This book was made with Forest
Stewardship Council™ certified
paper - one small step in DK's
commitment to a sustainable future.
**For more information go to
www.dk.com/our-green-pledge**

CONTENTS

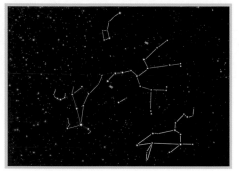

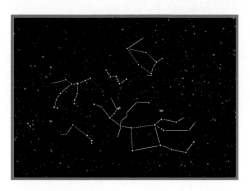

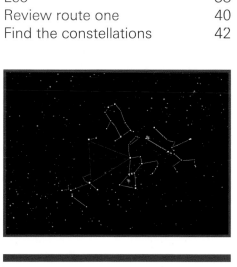

FOREWORD

▶▶▶ BY DR. MAGGIE ADERIN-POCOCK, MBE

On a clear night in the northern hemisphere, you can see up to 3,000 stars with the naked eye alone. When you look up at the stars, you are following a tradition that goes back thousands of years. As they tried to understand what they were seeing, people of ancient cultures looked up at the night sky and searched for patterns in the stars, eventually creating the ultimate dot-to-dot. They formed characters, animals, and objects from these patterns and made up stories about them that have passed down through generations.

By creating these patterns, called constellations, the stars were transformed from random dots into recognizable shapes that can be used to help us navigate our way through the night sky and down here on Earth.

With this book, you can follow in the footsteps of your ancestors by discovering the constellations and using them to hop from star to star and appreciate the beauty of the heavens.

Enjoy.

Maggie Aderin-Pocock

THE NIGHT SKY

YOU DO NOT HAVE TO MEMORIZE EVERY STAR IN ORDER TO APPRECIATE THE NIGHT SKY. INSTEAD, ASTRONOMERS LEARN SOME RECOGNIZABLE PATTERNS AND THEN FOLLOW STAR-HOPPING ROUTES ACROSS THE SKY.

A sea of stars
This stunning photograph reveals the thousands of stars that are visible in the night sky above Mono Lake in California, USA.

CONSTELLATIONS
PATTERNS IN THE NIGHT SKY

When you look up into the night sky on a clear night, you can see hundreds of seemingly **tiny pinpricks of light sprinkled across the sky**. For thousands of years, stargazers have looked for **patterns** among these lights, joining together the brightest ones to **form shapes and stories**.

1 In this patch of night sky, the larger dots represent the brightest stars and the smaller dots represent fainter stars. Thousands of years ago, astronomers began to make patterns from the brightest stars in the night sky.

2 The bright stars in this area of sky can be joined together to form the shape of a man. We call this shape an asterism. An asterism is any pattern of prominent stars, and the night sky is full of asterisms that have been recognized for millenia.

3 Astronomers imagined the patterns they found to be the people, gods, creatures, and objects that were told of in their stories. Ancient Greek astronomers decided that this pattern of a man represents the hunter Orion, who holds a club in one hand and the head of a lion in the other.

4 A constellation is an area of sky. The International Astronomical Union recognizes 88 altogether. Every star that lies within this orange outline around Orion is part of the constellation Orion. The official boundaries of the constellations create a map of the sky, which is used by astronomers around the world.

12

THE CELESTIAL SPHERE
A SPHERE OF STARS

Earth is surrounded by **hundreds of thousands** of visible stars, galaxies, and other objects. To help astronomers to chart and pinpoint the location of these deep-sky objects, they are **imagined on the surface of a sphere** that envelops Earth. We call this the **celestial sphere**.

THE STARRY SPHERE

Except for the Sun and the planets within the Solar System that move across the sky, every known object in space has been assigned a position on the surface of the celestial sphere that is more or less fixed.

1 Each of the 88 constellations in the night sky is found on the celestial sphere. Their jigsaw-like shapes can be joined together to form the shape of the sphere.

2 The celestial sphere is imagined as a spherical shell that surrounds Earth.

3 Earth lies at the centre looking out at the objects, such as stars, that are imagined to lie on the celestial sphere.

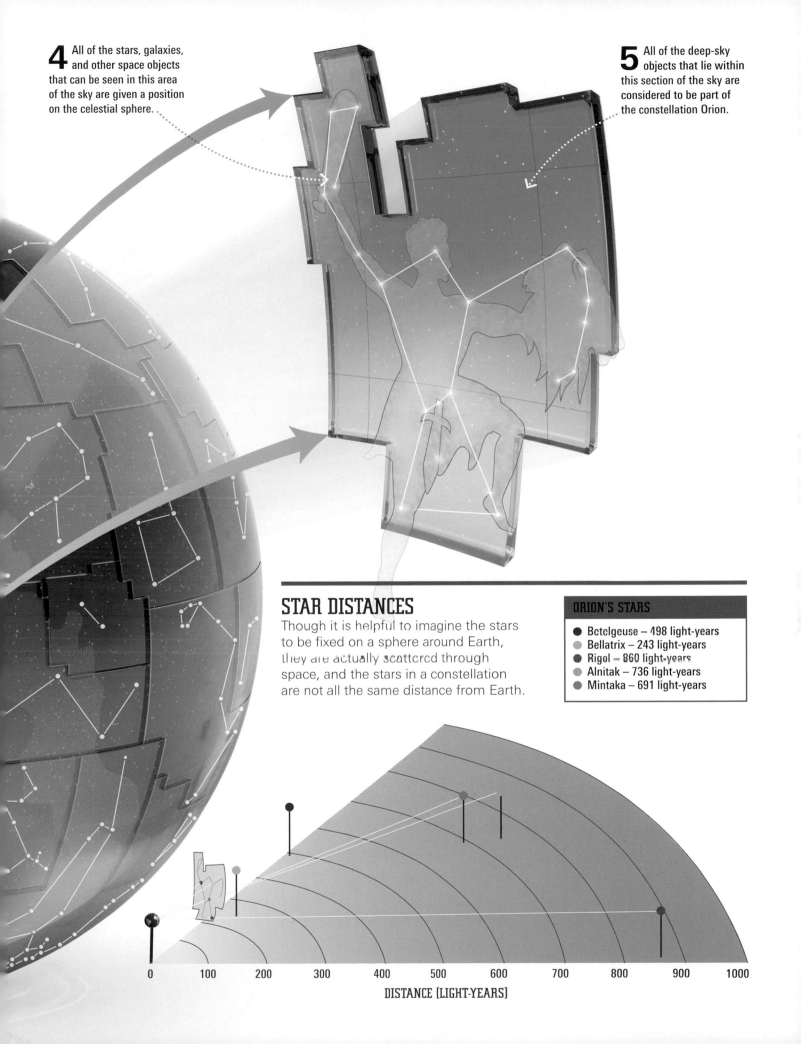

4 All of the stars, galaxies, and other space objects that can be seen in this area of the sky are given a position on the celestial sphere.

5 All of the deep-sky objects that lie within this section of the sky are considered to be part of the constellation Orion.

STAR DISTANCES

Though it is helpful to imagine the stars to be fixed on a sphere around Earth, they are actually scattered through space, and the stars in a constellation are not all the same distance from Earth.

ORION'S STARS

- Betelgeuse – 498 light-years
- Bellatrix – 243 light-years
- Rigel – 860 light-years
- Alnitak – 736 light-years
- Mintaka – 691 light-years

DISTANCE [LIGHT-YEARS]

0   100   200   300   400   500   600   700   800   900   1000

14

STAR MOVEMENT
OUR VIEW OF THE STARS

You can stargaze all year round and will be able to see **different constellations** as Earth orbits the Sun and faces a different part of space over the year. **Earth spins as each day and night passes**, so the stars appear to move across the sky **from east to west** throughout the night.

▲ This time-lapse photograph traces the movement of the stars over the course of an hour. The stars seem to rotate around Polaris, a nearly fixed point above the North Pole.

YOUR VIEW

Where you stand on Earth affects your view of the path of the stars. This is because the stars appear to rotate around the north celestial pole, a point that sits directly above Earth's North Pole.

1 At the North Pole, the north celestial pole is overhead, so stars circle around directly overhead.

2 At mid-latitudes, the north celestial pole is a distant point in the northern sky, so stars cross at an angle.

3 At the equator, the north celestial pole lies on the horizon, so stars cross the sky from east to west.

The best time to stargaze is on **a clear, dark night**. Your location will affect how many stars you can see. An open space, like a field, lets you **view more of the sky**. Places with bright lights can make it difficult to find some stars. The less light there is at your location, the more stars you will be able to see.

STARGAZING TIPS
ADVICE FOR STARGAZERS

1 Cities have bright lights, which give the sky a hazy glow called light pollution. In a city sky you will only see the brightest stars, but by looking carefully you can find some familiar asterisms.

2 Suburban towns have less light pollution than cities, so you will be able to see a few more stars and can start picking out the constellations.

3 Rural areas have very little light pollution and are good spots for stargazing. You will be able to see lots of stars and many of the constellations.

4 Dark-sky locations are far from any light pollution and are the best places to stargaze. You will be able to see the constellations, thousands of stars, and the band of our galaxy, the Milky Way.

STARGAZING TIPS

 Light pollution
Get as far away from artificial light as possible and try to find an open space to get a larger view of the sky.

Lunar phases
Find out the phase of the Moon – a full moon gives off so much light that it will be tricky to see the stars.

 Weather
Check the weather because thick cloud will block your view of the stars. The air cools down quickly at night, so dress in warm layers to stay comfortable while stargazing.

 Adjust to the dark
It can take at least 10 minutes for your eyes to adjust to the darkness, so be patient. Use a red light to see your star charts, because red light doesn't disturb your night vision like white light does.

 Seeing further
The naked eye can identify patterns in the stars, but a pair of binoculars or a telescope will enhance the detail of the night sky and allow you to find amazing sights such as double stars, galaxies, and nebulas.

THE PLOUGH
TO LEO

STARHOP FROM THE PLOUGH TO A FURTHER SEVEN PATTERNS IN THE NIGHT SKY, INCLUDING POLARIS (THE NORTHERN POLE STAR), AND SPOT TWO GALAXIES ALONG THE WAY. SPRING IS THE BEST TIME TO VIEW THIS ROUTE.

THE PLOUGH

URSA MAJOR

POLARIS

URSA MINOR

BOÖTES

CANES VENATICI

CORONA BOREALIS

LEO

The Plough
The most recognizable pattern in the night sky, the Plough (top, centre in this photograph), is where you will begin route one.

IF YOU LOOK NORTH, YOU WILL BE ABLE TO SEE THE SEVEN BRIGHT STARS THAT MAKE UP THE PLOUGH.

THE PLOUGH
THE BIG DIPPER

The **Plough** is a shape called an **asterism**. It makes up part of the large constellation Ursa Major. In the US, the Plough is known as the **Big Dipper**.

Double stars

A double star is a pair of stars that look very close together in the sky. However, the two stars may not actually be near each other in space. The Plough has a double star made up of the stars Mizar and Alcor, which appear right next to each other but are actually trillions of kilometres apart.

THE STAR ALCOR'S ORIGINAL ARABIC NAME MEANT "THE FORGOTTEN ONE"

MIZAR AND ALCOR

THE PLOUGH

This group of stars makes the shape of a simple **farmer's plough**, but is perhaps easier to recognize as a saucepan. The Plough is found by **looking north**.

THE PLOUGH

YOUR ROUTE ACROSS THE SKY

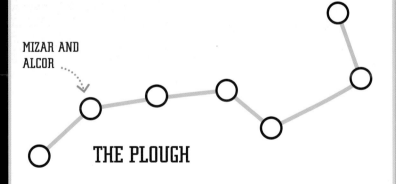

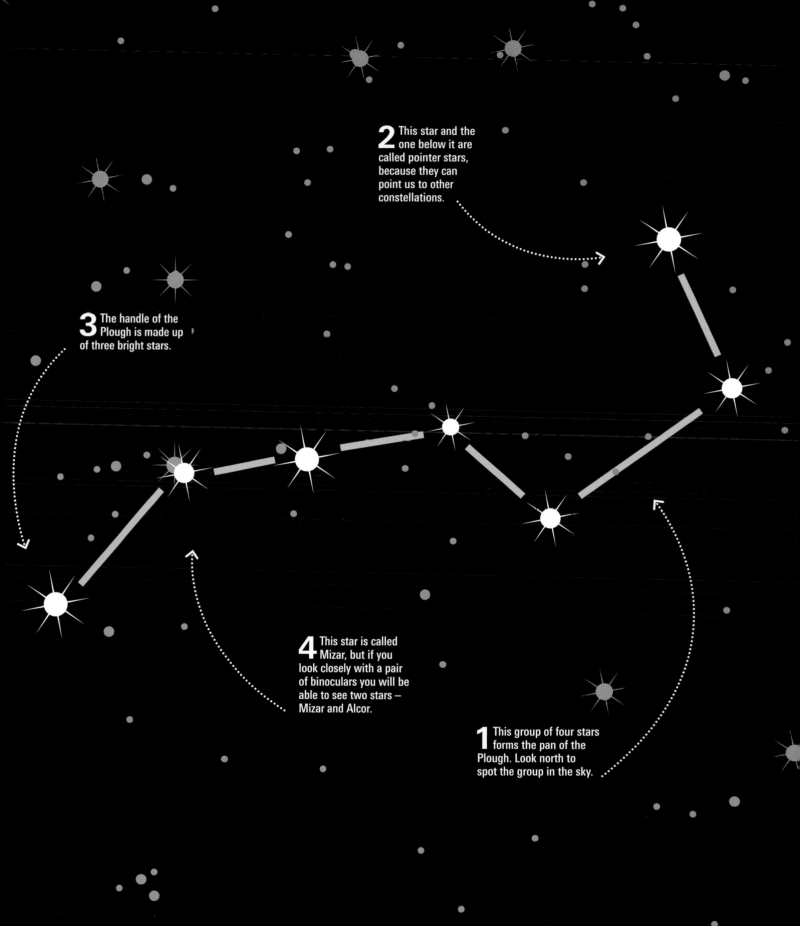

2 This star and the one below it are called pointer stars, because they can point us to other constellations.

3 The handle of the Plough is made up of three bright stars.

4 This star is called Mizar, but if you look closely with a pair of binoculars you will be able to see two stars — Mizar and Alcor.

1 This group of four stars forms the pan of the Plough. Look north to spot the group in the sky.

URSA MAJOR IS FOUND BY LOOKING FOR THE BRIGHT STARS THAT BRANCH OUT FROM THE PLOUGH.

URSA MAJOR
THE GREAT BEAR

Ursa Major means "**great bear**" and is the name we use to describe the **third largest constellation** in the sky. The Plough forms part of the constellation, with branches of bright stars making up the rest of the bear's shape.

Zeus's nymphs
Ursa Major represents the nymph Adrastea from Greek mythology. Zeus placed Adrastea and a nymph named Ida in the sky among the stars as the great bear and the little bear.

ANCIENT GREEKS USED URSA MAJOR TO HELP THEM NAVIGATE THE OCEANS AT NIGHT

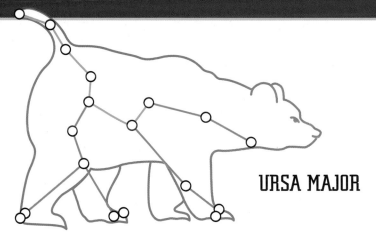

URSA MAJOR

The stars of **Ursa Major** can be linked to make the shape of a bear. **The Plough** makes up the body and tail, with **stars branching out** that make up the legs. The stars in the sky rotate around the north celestial pole throughout the night, so the constellation will not always appear this way up.

YOUR ROUTE ACROSS THE SKY

THE PLOUGH

URSA MAJOR

2 The two bright stars that branch off from the top of the pan form the head of the great bear.

3 The three bright stars that branch off from the bottom of the pan form the bear's front leg.

1 To find Ursa Major, first locate the Plough.

4 If you look for the bright stars below the back of the pan, you will find the bear's back legs.

5 In this view, the bear is tipped at an angle. This is because the stars rotate in the sky through the night.

THE CIGAR GALAXY CAN BE FOUND ABOVE THE NECK OF URSA MAJOR USING A TELESCOPE.

CIGAR GALAXY
MESSIER 82

This **starburst galaxy** is one of **more than 100 objects** in the night sky catalogued by French astronomer **Charles Messier** in the late 18th century. His list was designed to map the fixed objects in the night sky **so that he could easily spot comets** among them.

▲ Visible and infrared light captured by the Hubble Space Telescope, shows clouds of red hydrogen blasting out from the centre of the Cigar Galaxy (Messier 82).

Finding the Cigar Galaxy
To spot the Cigar Galaxy, first locate Ursa Major then the branch of stars that form its neck. The galaxy looks like a smudge in the sky. Using a telescope, it can be spotted just above the middle star that marks the bear's neck.

CIGAR GALAXY

POLARIS IS EASILY FOUND BY TRACING A LINE FROM THE POINTER STARS IN THE PLOUGH.

POLARIS
THE NORTHERN POLE STAR

Polaris, also known as the **northern pole star or North Star**, sits almost directly above Earth's North Pole. As Earth spins, the stars appear to **rotate around Polaris**, but Polaris itself seems to stay in one place.

Finding north

For hundreds of years, navigators have looked to Polaris to help them on their travels. Because it lies above Earth's North Pole, travellers knew that heading towards Polaris would take them north.

IN 1,000 YEARS POLARIS WILL NO LONGER BE THE NORTHERN POLE STAR

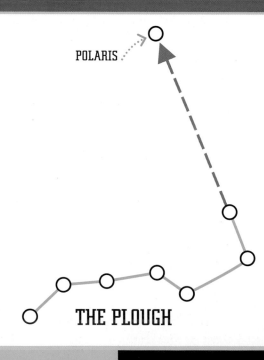

POLARIS

THE PLOUGH

Polaris is a **bright star in the northern sky** that is useful for navigation. You can find Polaris by by **tracing a line through the Plough's pointer stars** and following it until the first bright star you come to. The prominent star is part of a simple constellation called **Ursa Minor**.

YOUR ROUTE ACROSS THE SKY

THE PLOUGH

URSA MAJOR

POLARIS

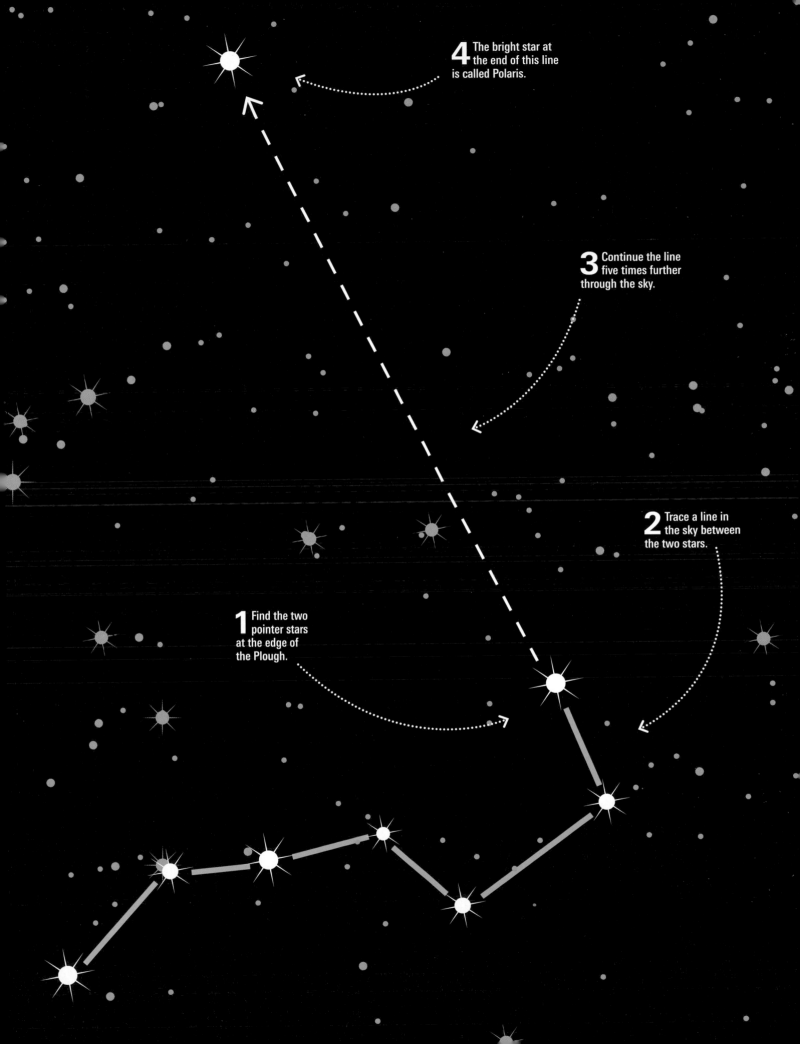

4 The bright star at the end of this line is called Polaris.

3 Continue the line five times further through the sky.

2 Trace a line in the sky between the two stars.

1 Find the two pointer stars at the edge of the Plough.

URSA MINOR CAN BE FOUND BY LOOKING FOR THE STARS THAT BRANCH OFF FROM POLARIS.

URSA MINOR
THE LITTLE BEAR

Ursa Minor is the closest constellation to the **north celestial pole** and includes the star Polaris. **You can always see Ursa Minor** in the northern sky, as it appears to spin around the northern pole star, Polaris.

Ida the nymph
According to Greek mythology, Ursa Minor represents a nymph named Ida. Alongside Adrastea depicted by Ursa Major, Ida nursed Zeus as an infant when he was hiding from his evil father. Zeus thanked the nymphs by transforming them into bears among the stars.

POLARIS IS A STAR THAT SHINES WITH THE BRIGHTNESS OF 2,500 SUNS

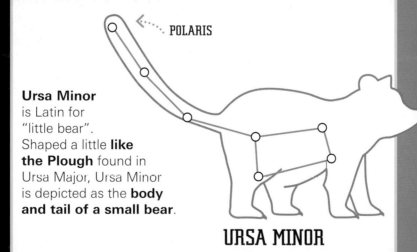

POLARIS

Ursa Minor is Latin for "little bear". Shaped a little **like the Plough** found in Ursa Major, Ursa Minor is depicted as the **body and tail of a small bear**.

URSA MINOR

YOUR ROUTE ACROSS THE SKY

THE PLOUGH

URSA MAJOR

POLARIS

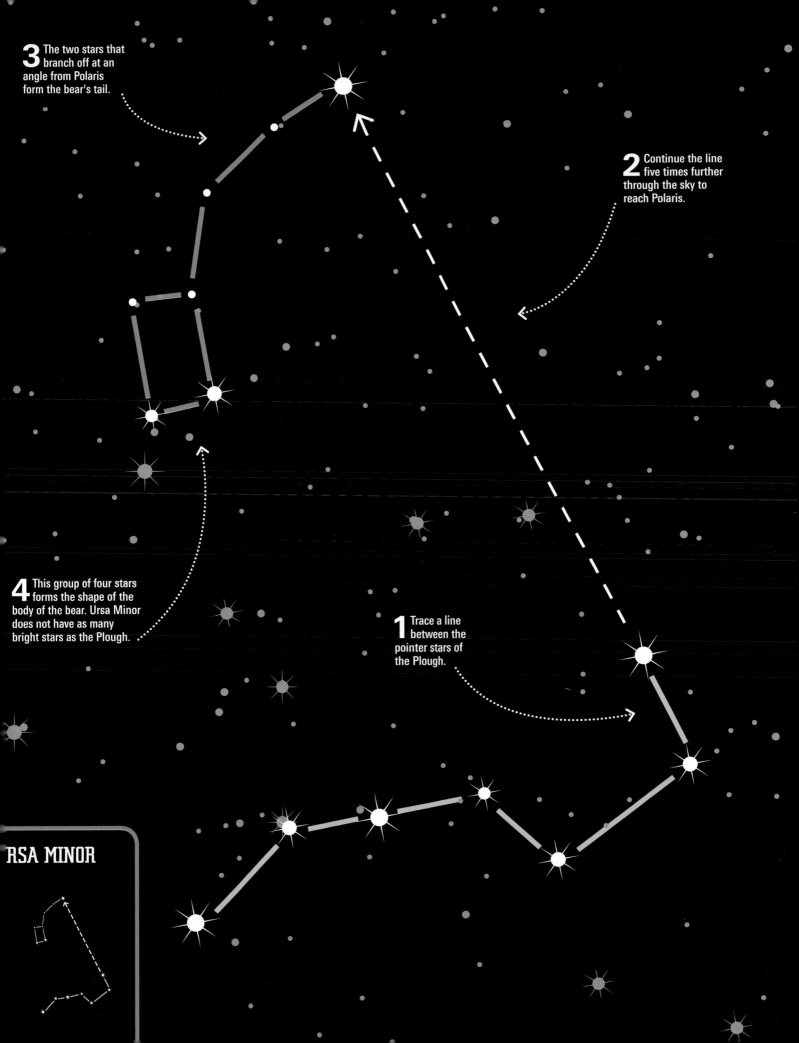

3 The two stars that branch off at an angle from Polaris form the bear's tail.

2 Continue the line five times further through the sky to reach Polaris.

4 This group of four stars forms the shape of the body of the bear. Ursa Minor does not have as many bright stars as the Plough.

1 Trace a line between the pointer stars of the Plough.

RSA MINOR

30

A SWEEPING ARC THROUGH THE SKY FROM THE HANDLE OF THE PLOUGH LEADS TO BOÖTES.

BOÖTES
THE HERDSMAN

Boötes contains one of the brightest stars in the night sky. The kite-shaped constellation represents a **herdsman** who chases the bears, **Ursa Major** and **Ursa Minor**, around the north celestial pole.

Boötes's planet
The faint star at the left knee of Boötes is called **Tau Boötis**. It is orbited by one of the **first planets discovered** beyond our Solar System, **Tau Boötis b**.

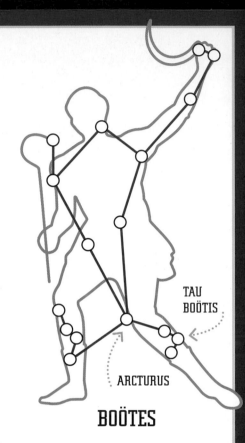

TAU BOÖTIS

ARCTURUS

BOÖTES

Boötes is depicted as **a man holding a staff** in one hand and a sickle in the other. A **kite-shaped** group of stars makes up the herdsman's body.

ARCTURUS RELEASES 100 TIMES MORE ENERGY THAN THE SUN

YOUR ROUTE ACROSS THE SKY

THE PLOUGH

URSA MAJOR

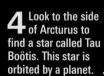

4 Look to the side of Arcturus to find a star called Tau Boötis. This star is orbited by a planet.

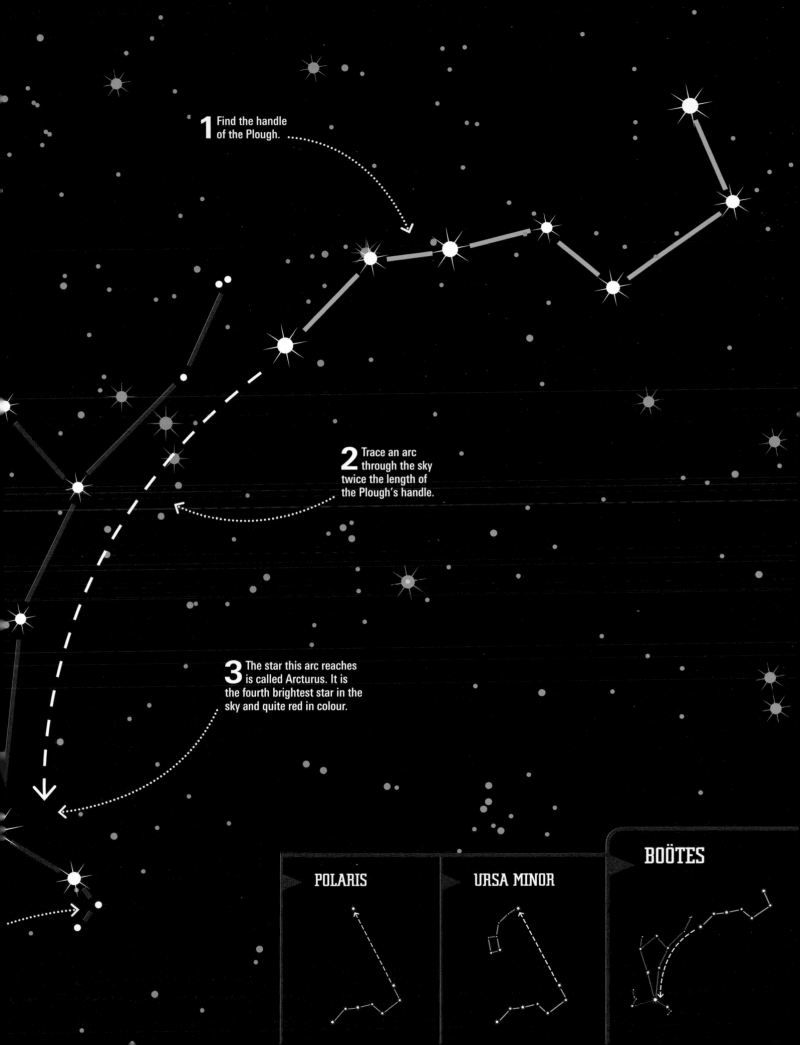

1 Find the handle of the Plough.

2 Trace an arc through the sky twice the length of the Plough's handle.

3 The star this arc reaches is called Arcturus. It is the fourth brightest star in the sky and quite red in colour.

POLARIS

URSA MINOR

BOÖTES

CANES VENATICI IS MADE UP OF TWO STARS THAT ARE FOUND TO THE SIDE OF BOÖTES.

CANES VENATICI
THE HUNTING DOGS

Lying between Boötes and Ursa Major, **Canes Venatici** represents Boötes's **hunting dogs**. They seem to **chase the bears**, Ursa Major and Ursa Minor, around the north celestial pole.

Heart of the king

Canes Venatici's brightest star is called **Cor Caroli**, which means "**Charles's heart**". It was named after an executed British king, Charles I.

THE MILKY WAY IS THOUGHT TO CONTAIN MORE THAN 100 BILLION STARS

2 Trace a line between the two stars.

Canes Venatici is a simple constellation made up of just **two stars** alongside Boötes. The constellation represents the herdsman's two **hunting dogs.**

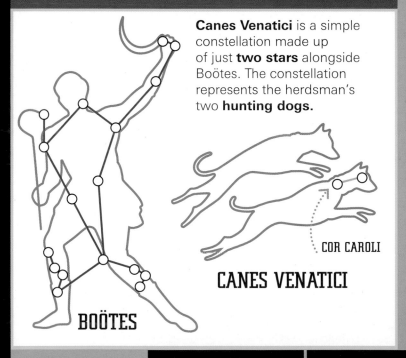

COR CAROLI

CANES VENATICI

BOÖTES

YOUR ROUTE ACROSS THE SKY

THE PLOUGH

URSA MAJOR

POLARIS

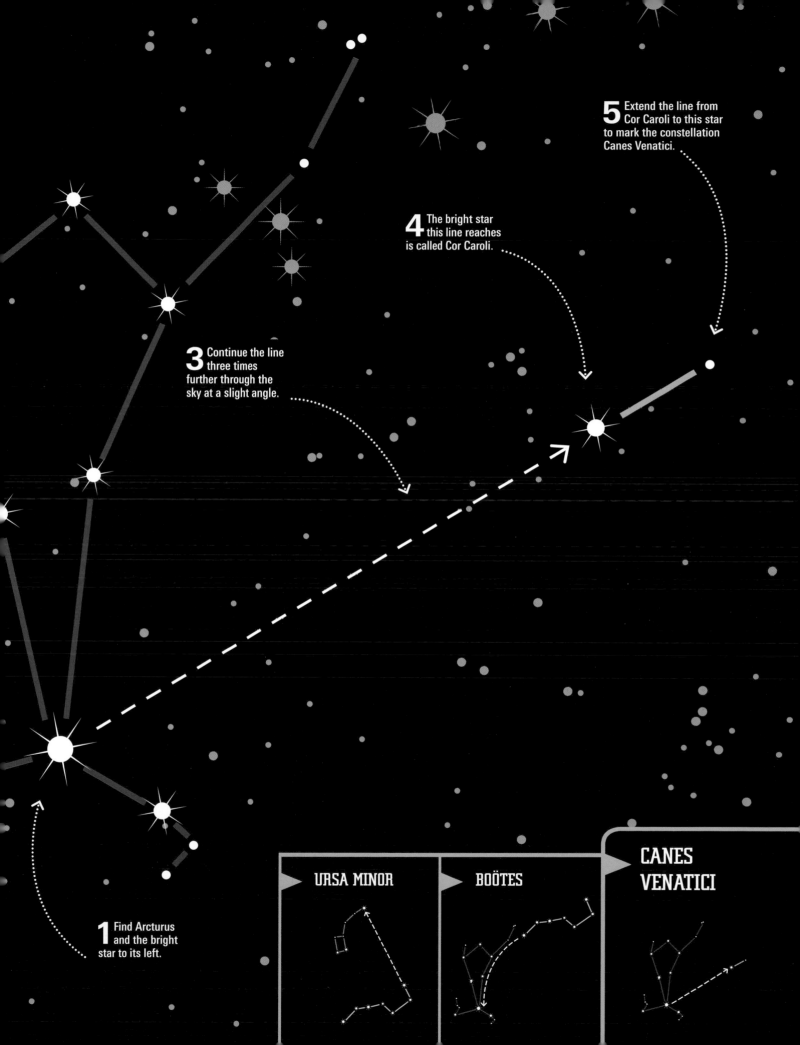

5 Extend the line from Cor Caroli to this star to mark the constellation Canes Venatici.

4 The bright star this line reaches is called Cor Caroli.

3 Continue the line three times further through the sky at a slight angle.

1 Find Arcturus and the bright star to its left.

URSA MINOR

BOÖTES

CANES VENATICI

THE WHIRLPOOL GALAXY LIES BETWEEN CANES VENATICI AND THE HANDLE OF THE PLOUGH.

WHIRLPOOL GALAXY

MESSIER 51

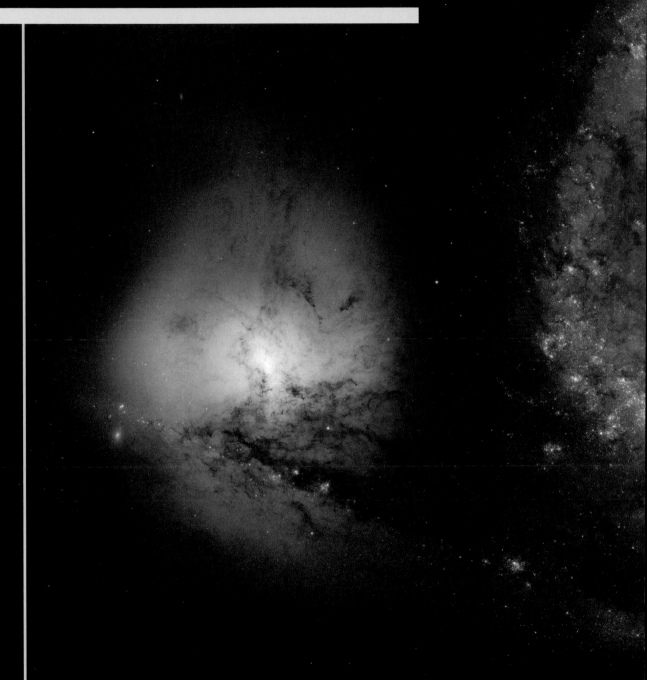

The **Hubble Space Telescope** captured this spectacular image of a vast, **sweeping spiral galaxy** connected by a delicate strand of gas to a smaller irregular galaxy. As it slowly passes behind the Whirlpool Galaxy, the irregular galaxy **triggers the birth of new stars**.

▲ Taken by the Hubble Space Telescope, this visible and infrared light image of the Whirlpool Galaxy shows a small yellow galaxy (named NGC 5195) at the tip of the larger Whirlpool Galaxy (also known as Messier 51).

Finding the Whirlpool Galaxy

The the easiest way to spot the Whirlpool Galaxy is by looking for the Plough. Using binoculars or a telescope, the galaxy can be seen just below the tip of the handle and looks like a smudge in the night sky.

WHIRLPOOL
GALAXY

CORONA BOREALIS IS MADE UP OF A SIMPLE ARC OF SEVEN STARS TO THE SIDE OF BOÖTES.

CORONA BOREALIS
THE NORTHERN CROWN

The constellation **Corona Borealis** is one of the original 48 constellations recognized in ancient Greece. It represents the **beautiful wedding crown** worn by the mythical Princess Ariadne of Crete.

Jewels in the sky

After his wedding to Princess Ariadne, the god Dionysus threw Ariadne's crown into the sky so its stunning jewels could be preserved forever as stars.

ON A CLEAR NIGHT, AS MANY AS 3,000 STARS CAN BE SEEN IN THE SKY BY THE NAKED EYE ALONE

The seven stars of **Corona Borealis** form a distinctive **horseshoe-shape of stars** in the night sky alongside the constellation Boötes. Each star in the constellation represents a **jewel in Ariadne's crown**.

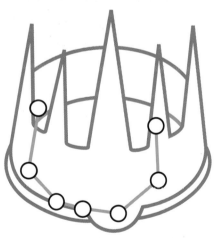

CORONA BOREALIS

YOUR ROUTE ACROSS THE SKY

THE PLOUGH

URSA MAJOR

POLARIS

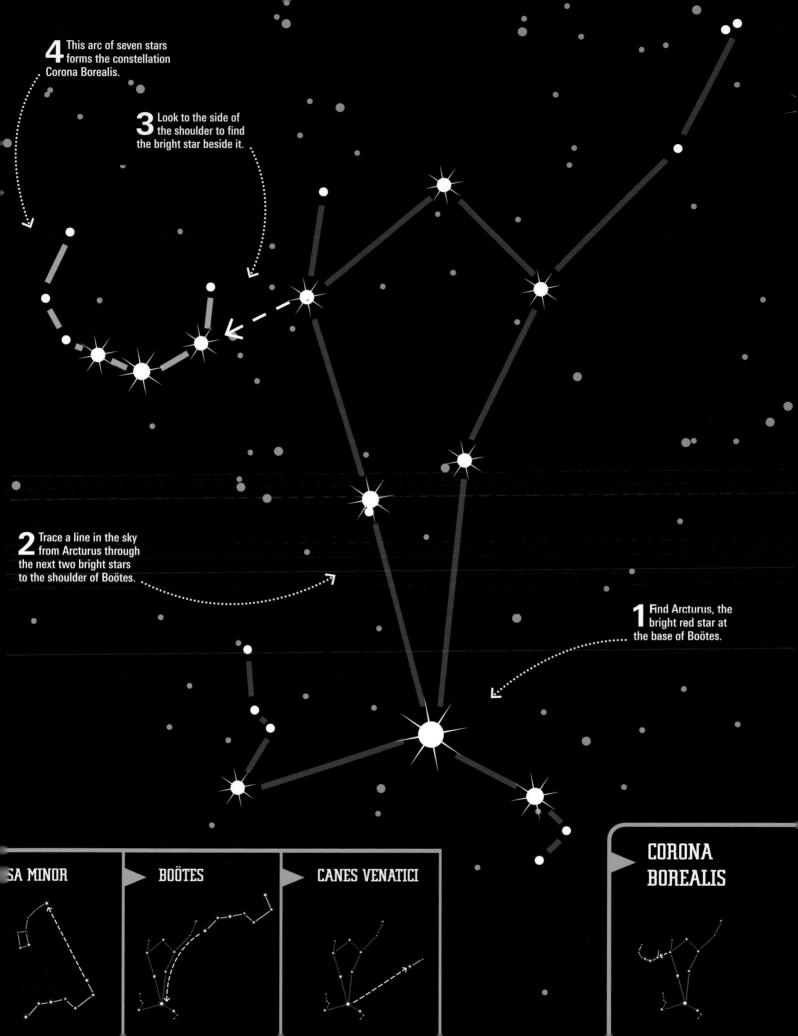

4 This arc of seven stars forms the constellation Corona Borealis.

3 Look to the side of the shoulder to find the bright star beside it.

2 Trace a line in the sky from Arcturus through the next two bright stars to the shoulder of Boötes.

1 Find Arcturus, the bright red star at the base of Boötes.

SA MINOR

BOÖTES

CANES VENATICI

CORONA BOREALIS

LEO CAN BE FOUND BY TRACING A LINE DOWN FROM THE POINTER STARS IN THE PLOUGH.

LEO
THE LION

Leo is a **zodiac constellation** and that represents the **lion** slain by Heracles of Greek mythology. **Heracles** wrestled with and defeated the lion as the first of his **12 labours**.

The zodiac
As Earth orbits the Sun each year, the Sun appears to pass in front of a band of sky where 12 constellations lie. We call these constellations the zodiac constellations.

REGULUS, LEO'S BRIGHTEST STAR, IS 79 LIGHT-YEARS AWAY FROM EARTH. THAT'S MORE THAN 747 TRILLION KM (464 TRILLION MILES) AWAY.

4 This group of stars represents Leo's body.

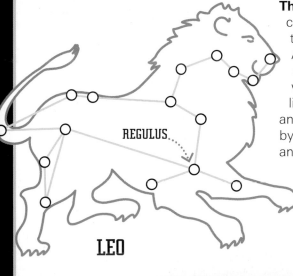

REGULUS

LEO

The stars of Leo can be joined to form the shape of **a lion**. A group of stars represents the body, with branches for the lion's legs. Leo's neck and chest are marked by **a string of six stars, an asterism called the Sickle**. An extra two stars represent the nose and jaw.

YOUR ROUTE ACROSS THE SKY

THE PLOUGH   URSA MAJOR   POLARIS

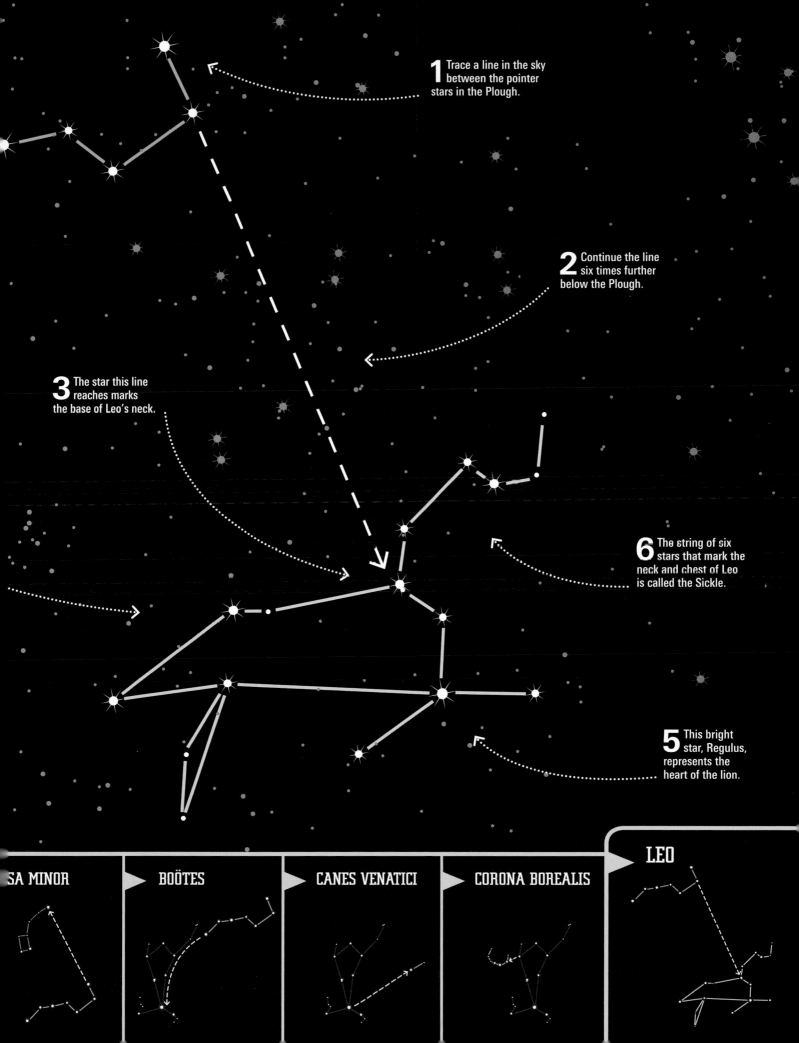

1 Trace a line in the sky between the pointer stars in the Plough.

2 Continue the line six times further below the Plough.

3 The star this line reaches marks the base of Leo's neck.

6 The string of six stars that mark the neck and chest of Leo is called the Sickle.

5 This bright star, Regulus, represents the heart of the lion.

SA MINOR → BOÖTES → CANES VENATICI → CORONA BOREALIS → **LEO**

40

THE CONSTELLATIONS IN ROUTE ONE CAN BE FOUND CLOSE TOGETHER IN THE SPRING SKY.

REVIEW ROUTE ONE
THE PLOUGH TO LEO

Here's what **the constellations of route one** look like when we **zoom out** to see them all together. The best time to look for these constellations is in **spring**, when they are found higher in the sky. At other times of year, when they are lower in the sky, you may not be able to see them.

CORONA BOREALIS

BOÖTES

ARCTURUS

TAU BOÖTIS

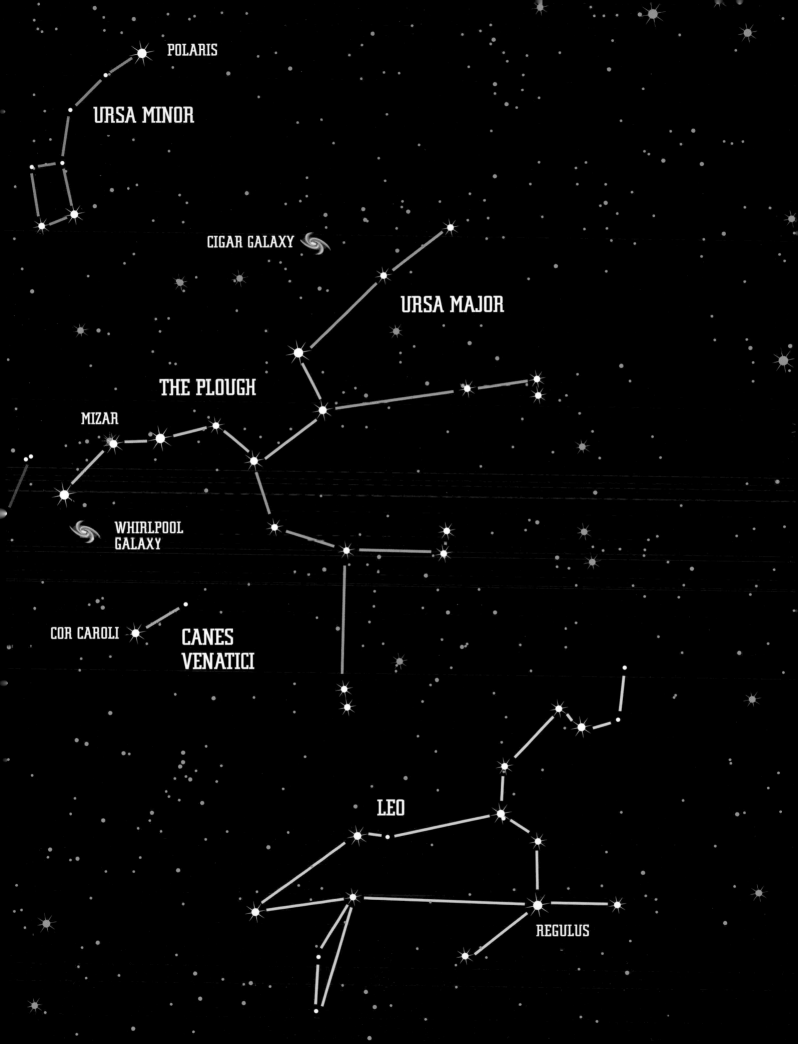

POLARIS

URSA MINOR

CIGAR GALAXY

URSA MAJOR

THE PLOUGH

MIZAR

WHIRLPOOL
GALAXY

COR CAROLI   CANES
VENATICI

LEO

REGULUS

CAN YOU USE THE ROUTE WE HAVE LEARNT TO NAVIGATE YOUR WAY THROUGH THIS NIGHT SKY?

FIND THE CONSTELLATIONS
THE PLOUGH TO LEO

Use the path you have learnt for route one to **practise finding your way** around this view of the night sky before heading outside to try it for yourself. Remember, the constellations **rotate around Polaris** through the evening, so may appear at different angles and in different areas of the sky.

YOUR ROUTE ACROSS THE SKY

| THE PLOUGH | URSA MAJOR | POLARIS |

SA MINOR > BOÖTES > CANES VENATICI > CORONA BOREALIS > LEO

ORION TO THE PLEIADES

FROM THE FAMOUS CONSTELLATION ORION, STARHOP BETWEEN SEVEN PATTERNS OF STARS, INCLUDING THE BRIGHTEST STAR IN THE NIGHT SKY AND A STAR CLUSTER CALLED THE PLEIADES. FOLLOW THIS ROUTE IN WINTER.

- ORION
- CANIS MAJOR
- CANIS MINOR
- WINTER TRIANGLE
- GEMINI
- TAURUS
- THE PLEIADES

Orion rising
This image shows Orion (centre) coming into view in New Mexico, USA. When a constellation rises, it is seen to move up from a spot low on the horizon.

FIND THE CONSTELLATION ORION BY LOOKING FOR THE THREE BRIGHT STARS THAT MAKE UP HIS BELT.

ORION
THE HUNTER

Holding a **club of bronze** in one hand and the head of a lion in the other, **Orion** represents a **mythical hunter**. Depicted facing the charging bull, Taurus, Orion is one of the most recognizable constellations in the **winter night sky**.

The mighty hunter
Son of the god Poseidon, Orion was a great hunter who slayed many ferocious beasts. After boasting that he could slay every beast on Earth, Orion was killed by a giant scorpion, which became the constellation Scorpius.

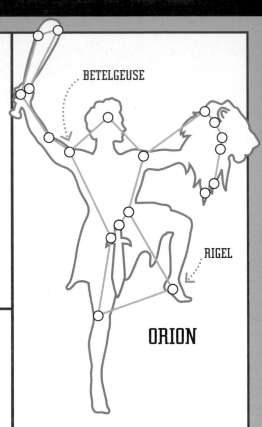

BETELGEUSE

RIGEL

ORION

The stars of **Orion** can be joined to form the shape of a **hunter**. A line of three stars at the centre makes up **Orion's Belt**, with stars branching off to create his body.

THE BRIGHT RED STAR BETELGEUSE IS ABOUT 1,000 TIMES THE SIZE OF THE SUN

ORION

YOUR
ROUTE
ACROSS
THE SKY

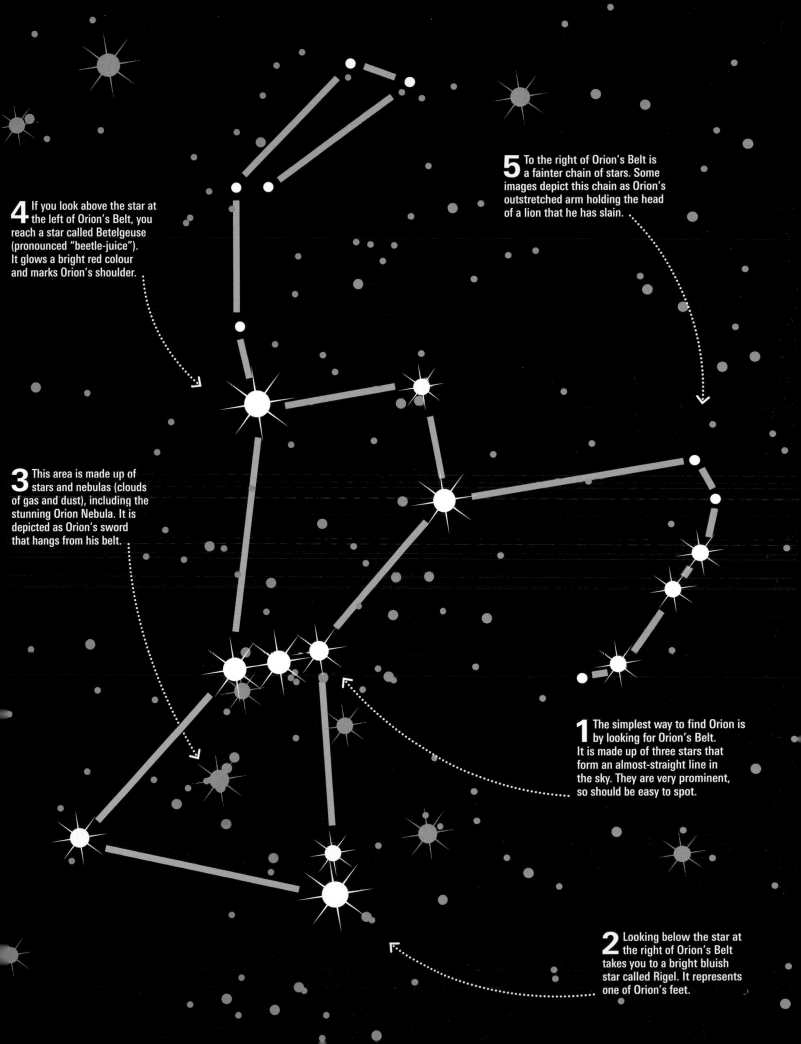

5 To the right of Orion's Belt is a fainter chain of stars. Some images depict this chain as Orion's outstretched arm holding the head of a lion that he has slain.

4 If you look above the star at the left of Orion's Belt, you reach a star called Betelgeuse (pronounced "beetle-juice"). It glows a bright red colour and marks Orion's shoulder.

3 This area is made up of stars and nebulas (clouds of gas and dust), including the stunning Orion Nebula. It is depicted as Orion's sword that hangs from his belt.

1 The simplest way to find Orion is by looking for Orion's Belt. It is made up of three stars that form an almost-straight line in the sky. They are very prominent, so should be easy to spot.

2 Looking below the star at the right of Orion's Belt takes you to a bright bluish star called Rigel. It represents one of Orion's feet.

THE ORION NEBULA LIES BENEATH ORION'S BELT AND MAKES UP PART OF HIS SWORD.

ORION NEBULA
MESSIER 42

Lying 1,300 light-years away from Earth, the Orion Nebula is an **immense cloud of gas and dust** in the constellation Orion. It is one of the brightest nebulas in the night sky, a place where **thousands of baby stars are being formed** as the gas and dust clouds within it collapse.

▲ This image, captured in 2016 by the HAWK-I infrared camera on the Very Large Telescope in Chile, reveals the Orion Nebula (Messier 42) in more detail than ever before.

Finding the Orion Nebula

To spot the Orion Nebula, look for the large, bright smudge that appears below Orion's Belt. This is the Orion Nebula and it represents part of Orion's sword. It appears as a smudge to the naked eye, but binoculars or a telescope will begin to reveal magnificent clouds of gas and dust.

ORION
NEBULA

CANIS MAJOR
THE GREATER DOG

One of the constellations identified by the ancient Greeks, **Canis Major** (pronounced can-iss may-jer) represents one of Orion's **hunting dogs**. The constellation contains **Sirius**, the **brightest star in the night sky**.

Laelaps
Canis Major represents Laelaps of Greek mythology, a dog so quick that no prey could escape him. When Laelaps failed to catch the Teumessian Fox, Zeus turned him to stone and placed him in the sky as Canis Major.

5 This star marks the dog's back, with stars extending out to make up its legs and tail.

THE NAME OF THE STAR SIRIUS COMES FROM A GREEK WORD MEANING "SCORCHING"

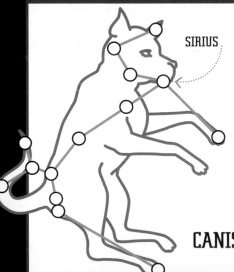

SIRIUS

The brightest star in the night sky, **Sirius**, represents the jaw of the dog **Laelaps**. Stars extend out from either side to mark the dog's ears and front leg. Below Sirius are several other **bright stars** that can be linked up to mark the dog's back, tail, and hind legs.

CANIS MAJOR

YOUR ROUTE ACROSS THE SKY

ORION

CANIS MAJOR

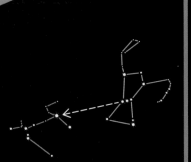

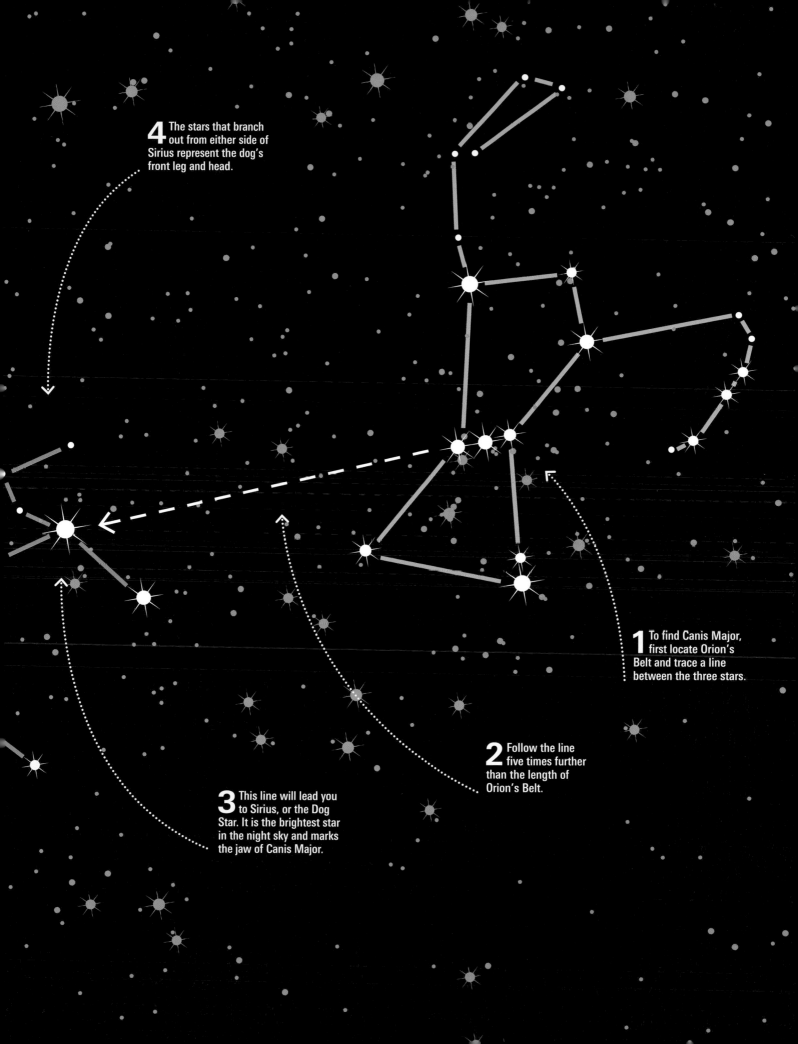

4 The stars that branch out from either side of Sirius represent the dog's front leg and head.

1 To find Canis Major, first locate Orion's Belt and trace a line between the three stars.

2 Follow the line five times further than the length of Orion's Belt.

3 This line will lead you to Sirius, or the Dog Star. It is the brightest star in the night sky and marks the jaw of Canis Major.

CANIS MINOR LIES ABOVE CANIS MAJOR AND JUST TO THE SIDE OF ORION'S RAISED HAND.

CANIS MINOR
THE LITTLE DOG

One of the smallest of the original Greek constellations, **Canis Minor** (pronounced can-iss my-ner) represents the smaller of Orion's two **hunting dogs**. The bright star **Procyon** makes it easy to pick out in the sky.

Early riser
The name Procyon means "before the dog". The star is so-called because it rises in the sky at night before the star Sirius, or the Dog Star, in Canis Major.

IN CHINESE ASTRONOMY, PROCYON FORMS PART OF A LARGE CONSTELLATION CALLED NANHE, THE SOUTHERN RIVER

5 These two stars can be joined together to make the constellation Canis Minor. The brighter of the two is called Procyon.

CANIS MINOR

PROCYON

Formed of just **two stars**, the pattern of **Canis Minor** is one of the **simplest** in the night sky. **Procyon**, the **brighter** of the two stars, represents the dog's body, while the other star marks the dog's neck.

YOUR ROUTE ACROSS THE SKY

ORION

CANIS MAJOR

CANIS MINO

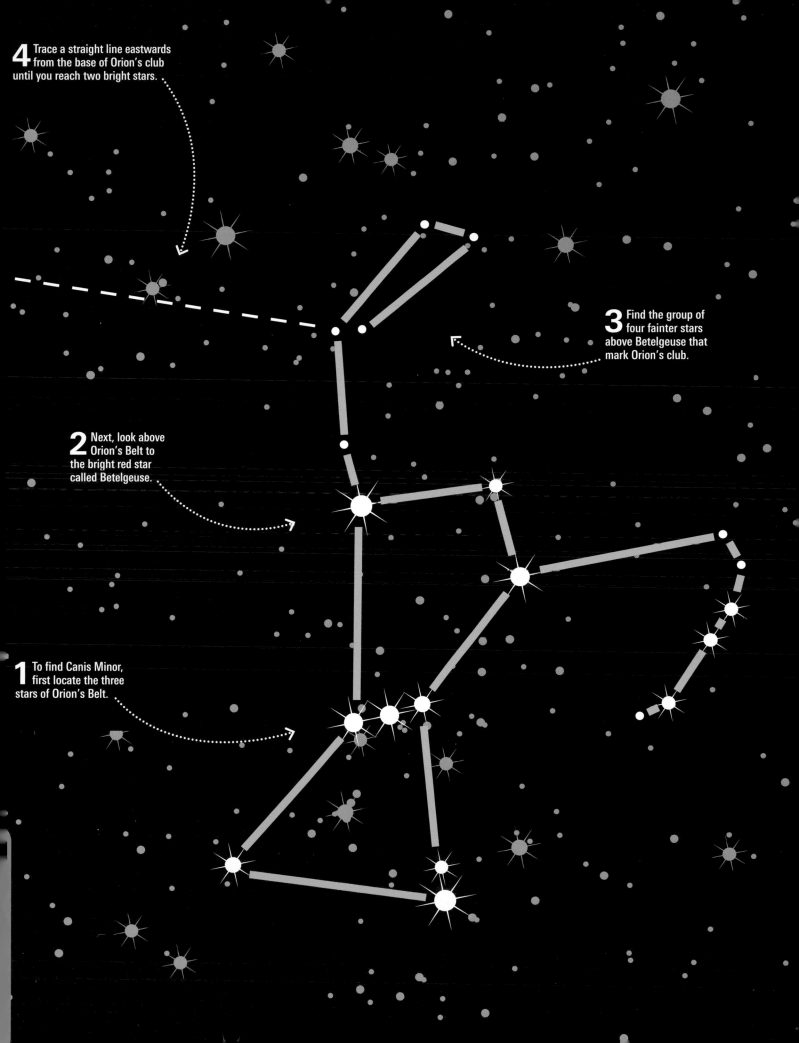

4 Trace a straight line eastwards from the base of Orion's club until you reach two bright stars.

3 Find the group of four fainter stars above Betelgeuse that mark Orion's club.

2 Next, look above Orion's Belt to the bright red star called Betelgeuse.

1 To find Canis Minor, first locate the three stars of Orion's Belt.

THE WINTER TRIANGLE IS MADE FROM THE BRIGHTEST STARS IN ORION, CANIS MAJOR, AND CANIS MINOR.

WINTER TRIANGLE
A WINTER ASTERISM

The Winter Triangle
is not a constellation, but a pattern of stars known as an **asterism**, consisting of stars from more than one constellation. It is formed of three stars – **Betelgeuse, Sirius, and Procyon**.

Light-years
Distances in space are so vast they are measured in a unit called light-years to make them easier to measure. One light-year is the distance that light can travel in one year, which is about 9.5 trillion km (5.9 trillion miles). Sirius is 8.6 light-years from Earth.

THE LIGHT EMITTED BY THE STAR SIRIUS TAKES 8.6 YEARS TO REACH EARTH

The shape of the **Winter Triangle** is a simple triangle. It has no story or picture like those that exist for other star patterns in the sky. Instead, it is just **three bright stars** from three prominent constellations.

PROCYON

BETELGEUSE

SIRIUS

WINTER TRIANGLE

YOUR ROUTE ACROSS THE SKY

ORION

CANIS MAJOR

CANIS MINOR

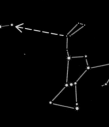

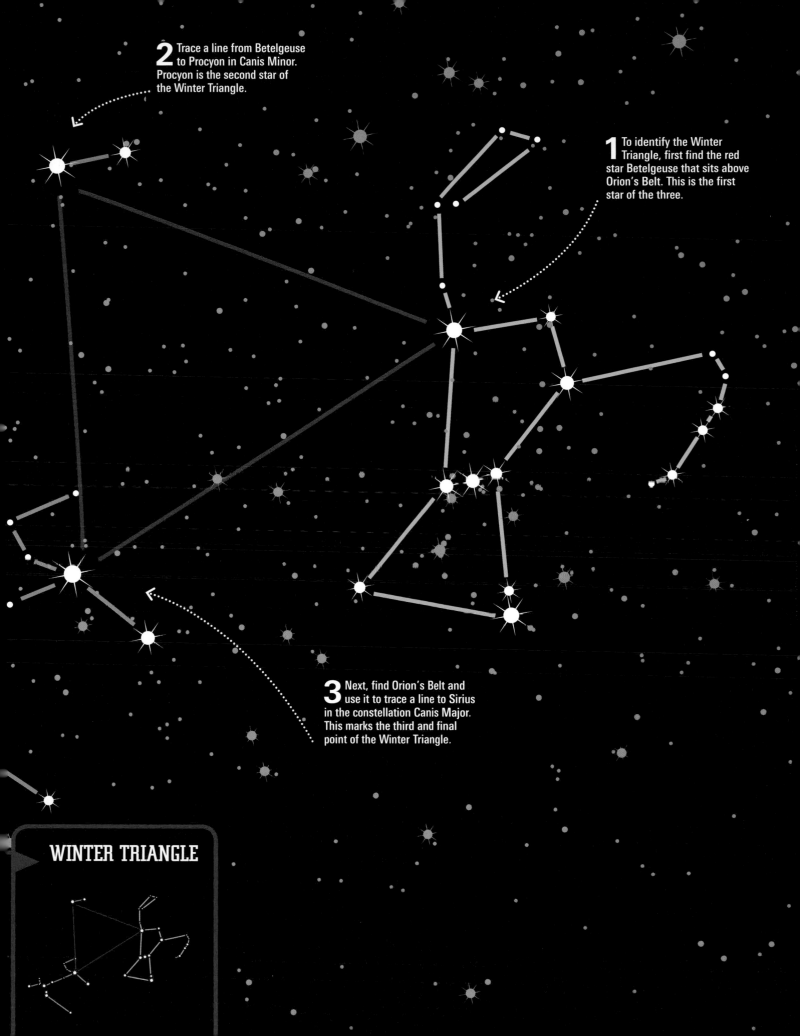

2 Trace a line from Betelgeuse to Procyon in Canis Minor. Procyon is the second star of the Winter Triangle.

1 To identify the Winter Triangle, first find the red star Betelgeuse that sits above Orion's Belt. This is the first star of the three.

3 Next, find Orion's Belt and use it to trace a line to Sirius in the constellation Canis Major. This marks the third and final point of the Winter Triangle.

WINTER TRIANGLE

ORION TO THE PLEIADES : GEMINI

56

GEMINI IS FOUND BY TRACING A LINE THROUGH
THE SKY FROM RIGEL AND BETELGEUSE IN ORION.

5 This star, called Pollux, represents the head of the twin Pollux.

GEMINI
THE TWINS

Gemini is one of the most recognizable of the constellations of the **zodiac**. It is pictured in the night sky as the **twins Castor and Pollux** of Greek mythology, sons of Queen Leda of Sparta.

The Castor system
While the star Castor looks like a solo star to the naked eye, using a telescope reveals a second star, bound to Castor by gravity. In fact, both of these stars are double stars and lie near another double star, making a group of six. So, when you look up at Castor, you are seeing a group of six stars, not just one star.

CASTOR AND POLLUX HAD THE POWER TO PROTECT SAILORS WHO WERE IN TROUBLE AT SEA

The stars of **Gemini** are depicted as the **twins Castor and Pollux**. The constellation is roughly rectangular in shape – its **two brightest stars** mark the twins' heads, with a chain of stars to depict their bodies and feet.

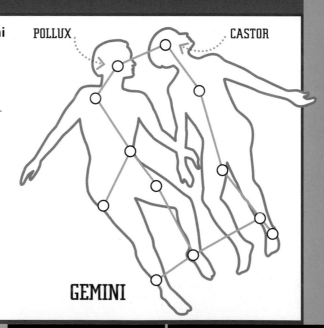

POLLUX   CASTOR

GEMINI

YOUR ROUTE ACROSS THE SKY

ORION

CANIS MAJOR

CANIS MINOR

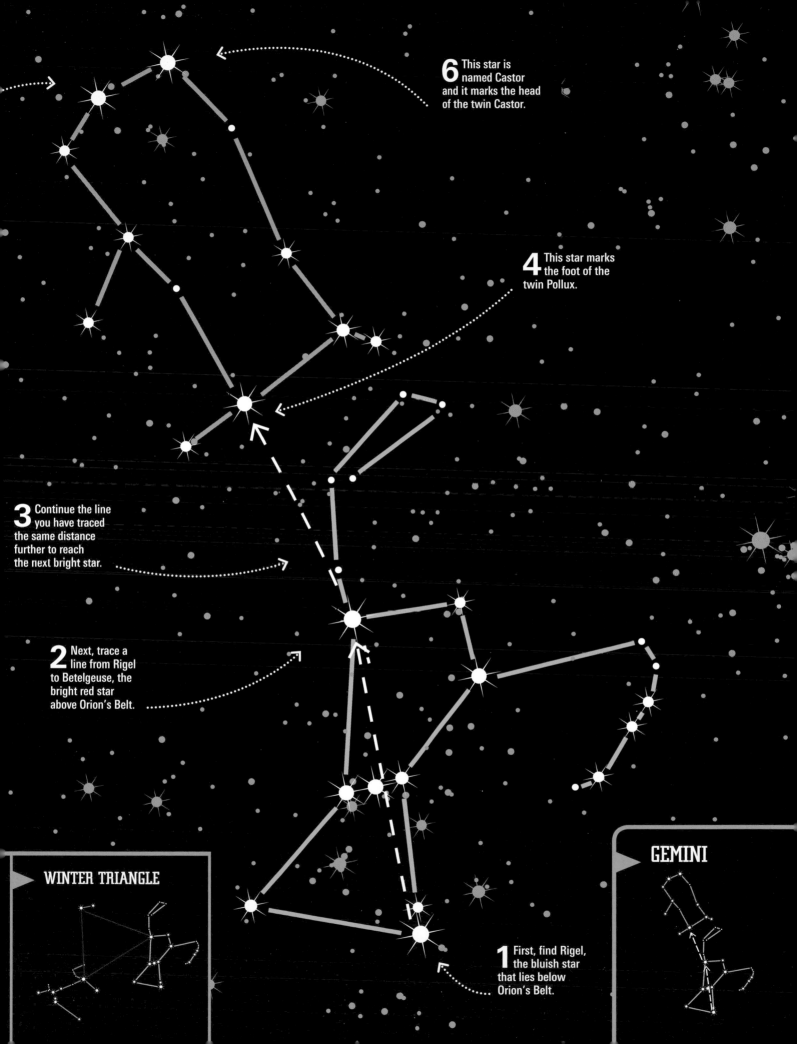

6 This star is named Castor and it marks the head of the twin Castor.

4 This star marks the foot of the twin Pollux.

3 Continue the line you have traced the same distance further to reach the next bright star.

2 Next, trace a line from Rigel to Betelgeuse, the bright red star above Orion's Belt.

1 First, find Rigel, the bluish star that lies below Orion's Belt.

WINTER TRIANGLE

GEMINI

58 FIND THE EYE OF TAURUS BY TRACING A CURVED LINE ALONG THE ARM OF ORION.

TAURUS THE BULL

The constellation Taurus depicts a **bull** charging at Orion. It has been recognized since Babylonian times, more than **2,500 years ago**. It is easily recognized by the V-shaped cluster that marks its head.

Open clusters

A star cluster is a group of stars. Some, called globular clusters, are tightly packed together, while others are grouped loosely to form an open cluster. The Hyades, the face of Taurus, is an open cluster containing about 200 stars.

ANCIENT GREEKS BELIEVED TAURUS WAS THE GOD ZEUS IN DISGUISE AS A BULL TO ATTRACT A MAIDEN

The stars of **Taurus** can be linked together to depict an **angry bull** that faces Orion. A prominent cluster of stars, called the **Hyades**, forms the V-shape that represents Taurus's head, with two bright stars above marking the **tips of his horns**. Two branches of stars from his nose form the bull's two legs.

TAURUS

ALDEBARAN

1 Find Orion by looking for the three stars marking his belt.

YOUR ROUTE ACROSS THE SKY

ORION

CANIS MAJOR

CANIS MINOR

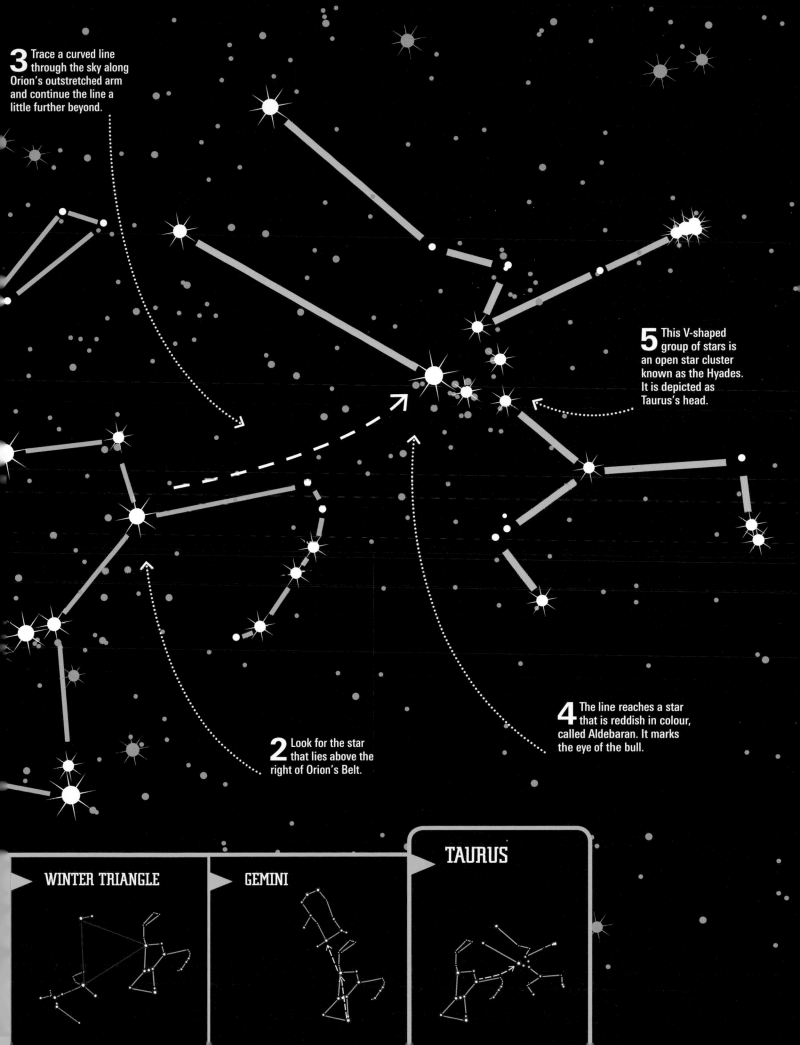

3 Trace a curved line through the sky along Orion's outstretched arm and continue the line a little further beyond.

5 This V-shaped group of stars is an open star cluster known as the Hyades. It is depicted as Taurus's head.

4 The line reaches a star that is reddish in colour, called Aldebaran. It marks the eye of the bull.

2 Look for the star that lies above the right of Orion's Belt.

WINTER TRIANGLE

GEMINI

TAURUS

60

THE CRAB NEBULA IS FOUND VERY CLOSE
TO THE TIP OF TAURUS'S LEFT HORN.

CRAB NEBULA
MESSIER 1

Lying within the constellation Taurus, the Crab
Nebula was formed in 1054 CE from the **enormous
explosion of a dying star**. At the centre lies the
core of the exploded star, called a **pulsar**. The cloud
of gas and dust it has left behind is known as a
supernova remnant.

▲ This image of the
Crab Nebula (Messier
1), captured by the
Hubble Space
Telescope, reveals the
strands of gas and dust
that have been ejected
by the stellar explosion.

Finding the Crab Nebula

Located just to the side of the tip of
Taurus's left horn, the Crab Nebula
appears as a faint blotch through
binoculars. A good telescope will
reveal some of the detail of the strands
that extend out from the centre.

CRAB
NEBULA

62

THE PLEIADES IS AN OPEN STAR CLUSTER REPRESENTING THE SHOULDER OF TAURUS.

THE PLEIADES
THE SEVEN SISTERS

The Pleiades
(pronounced play-a-dees) is a **100 million-year-old** open star cluster in **Taurus**. While six stars are easily visible to the naked eye, the Pleiades contains **hundreds of stars**.

The missing Pleiad
Only six of the seven sisters are easily seen in the Pleiades. One myth says that Merope, the youngest of the sisters, shines less brightly because she married a mortal, Sisyphus, rather than a god.

THE PLEIADES WERE THE SEVEN DIVINE DAUGHTERS OF ATLAS AND PLEIONE

The Pleiades is an open star cluster found within the constellation **Taurus**, marking the bull's shoulder. The cluster is made up of **six stars** that are visible to the naked eye, but binoculars reveal **nine bright stars** in the cluster – the **seven divine sisters** and their two parents.

THE PLEIADES

TAURUS

YOUR ROUTE ACROSS THE SKY

ORION

CANIS MAJOR

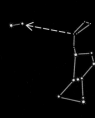

CANIS MINOR

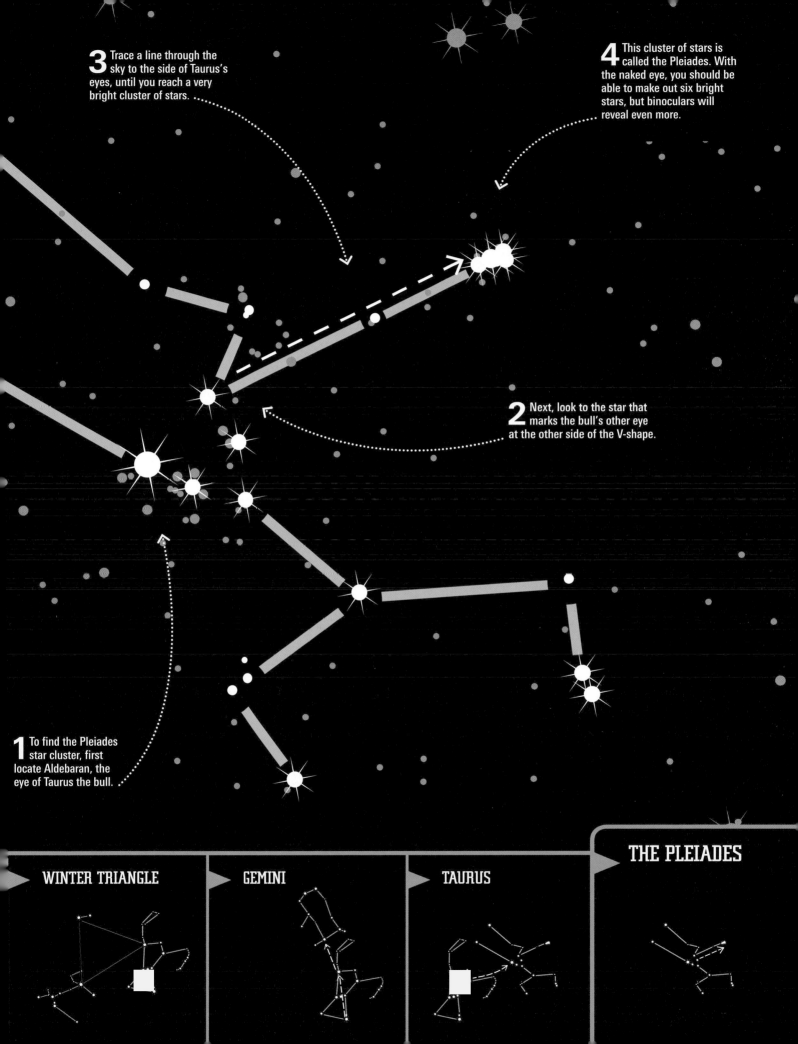

3 Trace a line through the sky to the side of Taurus's eyes, until you reach a very bright cluster of stars.

4 This cluster of stars is called the Pleiades. With the naked eye, you should be able to make out six bright stars, but binoculars will reveal even more.

2 Next, look to the star that marks the bull's other eye at the other side of the V-shape.

1 To find the Pleiades star cluster, first locate Aldebaran, the eye of Taurus the bull.

WINTER TRIANGLE

GEMINI

TAURUS

THE PLEIADES

STARGAZING IN WINTER MONTHS WILL REVEAL THE CONSTELLATIONS OF ROUTE TWO.

REVIEW ROUTE TWO
ORION TO THE PLEIADES

When we zoom out a little this is what all the constellations of **route two** look like. The best time to look for these constellations is during the **winter** when they appear **high in the sky**. At other times of the year, they are not visible in the night sky.

CANIS MINOR

PROCYON

WINTER TRIANGLE

CANIS MAJOR

SIRIUS

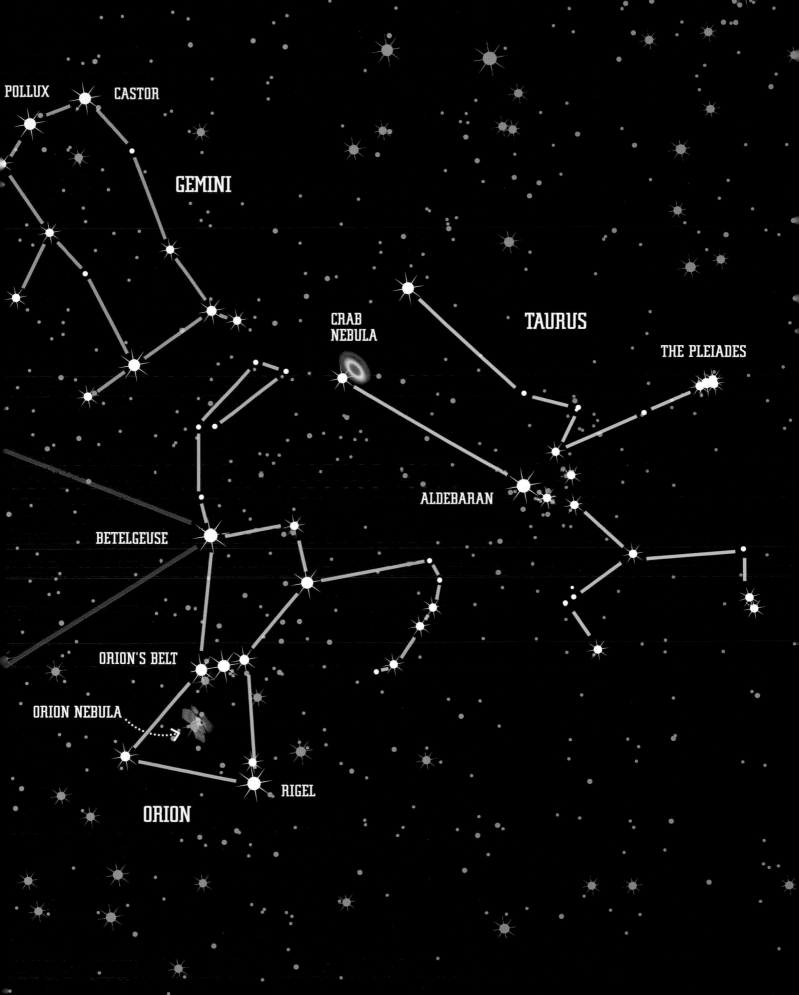

HAVE A GO AT FINDING YOUR WAY THROUGH THIS SKY USING THE ROUTE YOU HAVE LEARNT.

FIND THE CONSTELLATIONS
ORION TO THE PLEIADES

Here you can **practise route two** before heading outside to try it out with the **real night sky**. Remember, light pollution can make it tricky to see fainter stars, so find a **dark spot** for stargazing.

YOUR ROUTE ACROSS THE SKY

ORION

CANIS MAJOR

CANIS MINOR

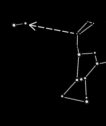

WINTER TRIANGLE

GEMINI

TAURUS

THE PLEIADES

CASSIOPEIA TO ARIES

STARHOP FROM CASSIOPEIA TO FIVE MORE SHAPES IN THE STARS, INCLUDING THE WINGED HORSE PEGASUS, AND SPOT OUR CLOSEST GALAXY, ANDROMEDA, ON THE WAY. VIEW THIS ROUTE IN AUTUMN.

- CASSIOPEIA
- CEPHEUS
- PERSEUS
- ANDROMEDA
- PEGASUS
- ARIES

La Palma telescopes
Cassiopeia (far right) is pictured above the Roque de Los Muchachos Observatory in the Canary Islands, home to some of the most powerful telescopes in the world.

CASSIOPEIA IS A RECOGNIZABLE PATTERN FOUND ON THE OPPOSITE SIDE OF POLARIS TO THE PLOUGH.

CASSIOPEIA
QUEEN OF ETHIOPIA

Representing Greek mythology's **Queen Cassiopeia of Ethiopia**, the constellation Cassiopeia (pronounced cass-ee-oh-pee-uh) has a distinctive **W-shape** of **five bright stars** that makes it easy to spot in the night sky.

A vain queen
Queen Cassiopeia was renowned for her vanity. Boasting of her beauty, she infuriated the Nereids, daughters of Poseidon. To punish her, Poseidon sent a terrifying sea monster, called Cetus, to destroy her kingdom.

GAMMA CASSIOPEIAE ROTATES AT MORE THAN 1 MILLION KM/H (625,000 MPH) AT ITS EQUATOR

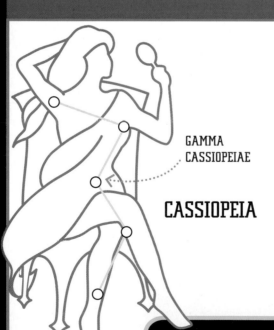

GAMMA CASSIOPEIAE

CASSIOPEIA

The constellation Cassiopeia does not much resemble a queen. Instead, the constellation's **five brightest stars** are linked to **form a W-shape**, where two stars marking her legs are joined to the two stars marking her shoulders by a star called **Gamma Cassiopeiae**.

CASSIOPEIA

YOUR ROUTE ACROSS THE SKY

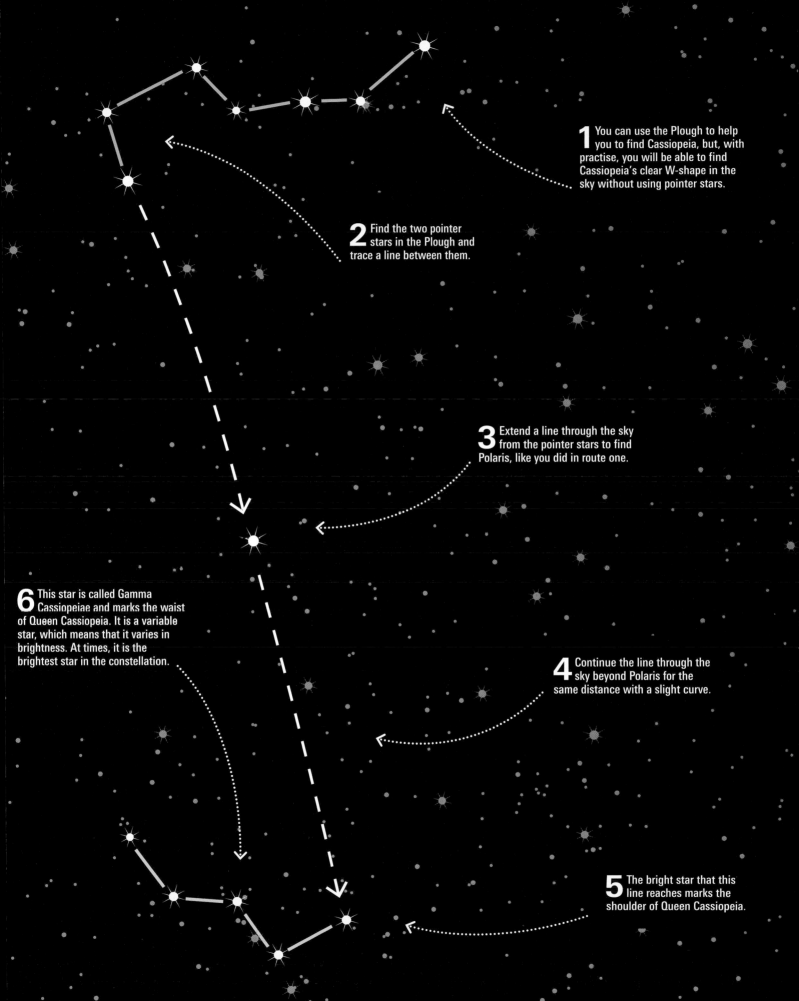

1 You can use the Plough to help you to find Cassiopeia, but, with practise, you will be able to find Cassiopeia's clear W-shape in the sky without using pointer stars.

2 Find the two pointer stars in the Plough and trace a line between them.

3 Extend a line through the sky from the pointer stars to find Polaris, like you did in route one.

6 This star is called Gamma Cassiopeiae and marks the waist of Queen Cassiopeia. It is a variable star, which means that it varies in brightness. At times, it is the brightest star in the constellation.

4 Continue the line through the sky beyond Polaris for the same distance with a slight curve.

5 The bright star that this line reaches marks the shoulder of Queen Cassiopeia.

FIND CEPHEUS BY TRACING A LINE FROM THE THREE BRIGHTEST STARS IN CASSIOPEIA.

CEPHEUS
KING OF ETHIOPIA

The constellation Cepheus (pronounced sef-ee-us) is depicted as the **mythical King Cepheus of Ethiopia**, husband of Queen Cassiopeia. Looking a little like a church in shape, Cepheus lies to the side of his queen.

Cepheid variables
Stars that vary in brightness are called Cepheid variables, because the first identified was the star Delta Cephei. Even at more than 800 light-years away, it can be seen to vary in brightness every five days by the naked eye.

2 Next, trace a line between these three bright stars.

CEPHEID VARIABLES PULSATE: THEY BRIGHTEN AND DIM REGULARLY OVER A PERIOD OF TIME

DELTA CEPHEI

Said to represent **King Cepheus of Ethiopia**, the constellation Cepheus is an unusual shape. The pattern of the constellation looks a little more like **the shape of a church and steeple** than the king. The point of the steeple marks the knee of Cepheus and the base represents his head and shoulders.

CEPHEUS

YOUR ROUTE ACROSS THE SKY

CASSIOPEIA

CEPHEUS

6 This star marks Cepheus's right knee.

5 Look for the shape of a church and steeple made out of bright stars to identify the rest of the constellation Cepheus.

4 This star is called Delta Cephei. Its brightness varies every five days.

3 Continue the line that you have traced with a gentle arc through the sky until you reach the next bright star. This star marks the head of Cepheus.

1 To find Cepheus, first look for the second, third, and fifth star in the W-shape of Cassiopeia.

PERSEUS CAN BE FOUND BY TRACING A LINE FROM CASSIOPEIA'S THREE BRIGHTEST STARS.

PERSEUS
THE VICTORIOUS HERO

Perseus is a bright constellation depicted as the hero from ancient Greek mythology. Known for his **heroic defeat** of the **snake-haired Gorgon Medusa**, Perseus is pictured in the sky holding the monstrous Medusa's severed head.

Andromeda's hero

On his return from defeating Medusa, Perseus saw Princess Andromeda, daughter of King Cepheus and Queen Cassiopeia, chained to a rock as a sacrifice to a sea monster. He killed the monster and freed Andromeda, taking her as his bride. They lie side-by-side in the sky.

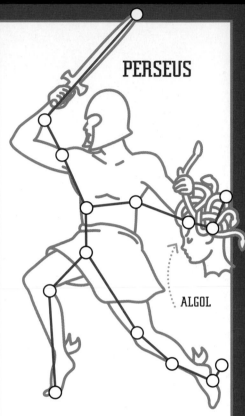

PERSEUS

ALGOL

5 Two simple branches of stars mark Perseus's legs.

Depicted as a **Greek hero**, the major stars of Perseus can be joined to form the two legs and two arms of a man. He holds a **sword high in one hand** and the **head of Medusa** in the other.

THE PERSEID METEOR SHOWER RADIATES FROM NORTHERN PERSEUS EACH AUGUST

YOUR ROUTE ACROSS THE SKY

CASSIOPEIA

CEPHEUS

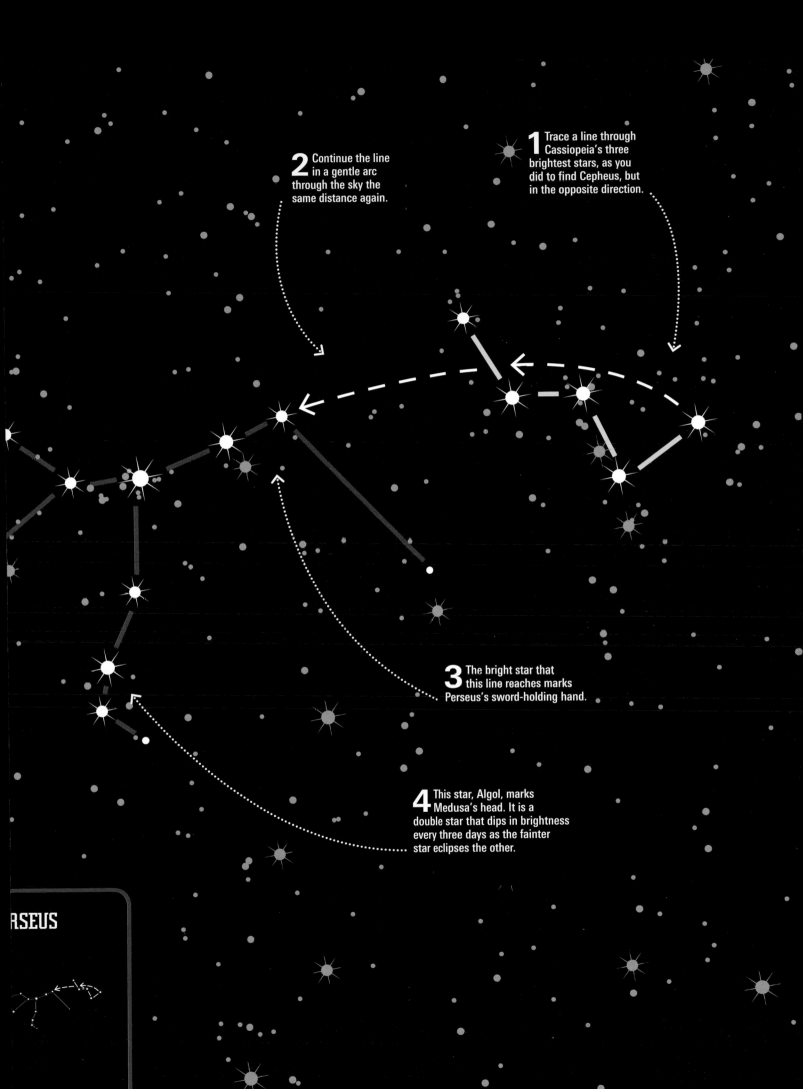

2 Continue the line in a gentle arc through the sky the same distance again.

1 Trace a line through Cassiopeia's three brightest stars, as you did to find Cepheus, but in the opposite direction.

3 The bright star that this line reaches marks Perseus's sword-holding hand.

4 This star, Algol, marks Medusa's head. It is a double star that dips in brightness every three days as the fainter star eclipses the other.

RSEUS

76

ANDROMEDA IS FOUND BY FOLLOWING THE END
OF PERSEUS'S SWORD TO ANDROMEDA'S HIP.

ANDROMEDA
THE CAPTIVE PRINCESS

Andromeda is a mythical princess, daughter of King Cepheus and Queen Cassiopeia of Ethiopia. When their kingdom was ravaged by the **sea monster Cetus**, the king and queen were ordered by the gods to **sacrifice their daughter** to appease the beast.

A shared star
The bright star Alpheratz, which marks Andromeda's head, is also part of an asterism in Pegasus called the Great Square of Pegasus. However, Alpheratz is officially designated as part of the Andromeda constellation only.

AFTER ANDROMEDA'S DEATH, THE GODDESS ATHENA PLACED HER IN THE SKY TO HONOUR HER

Similar in shape to the constellation Perseus, the bright stars of Andromeda can be joined to form the shape of the body of **Princess Andromeda**. Chains of stars branching out from the centre mark her arms, legs, and the **chains that restrained her** when she was offered as a sacrifice to the sea monster Cetus.

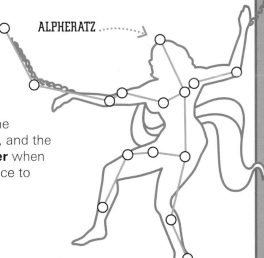

ALPHERATZ

ANDROMEDA

YOUR ROUTE ACROSS THE SKY

CASSIOPEIA

CEPHEUS

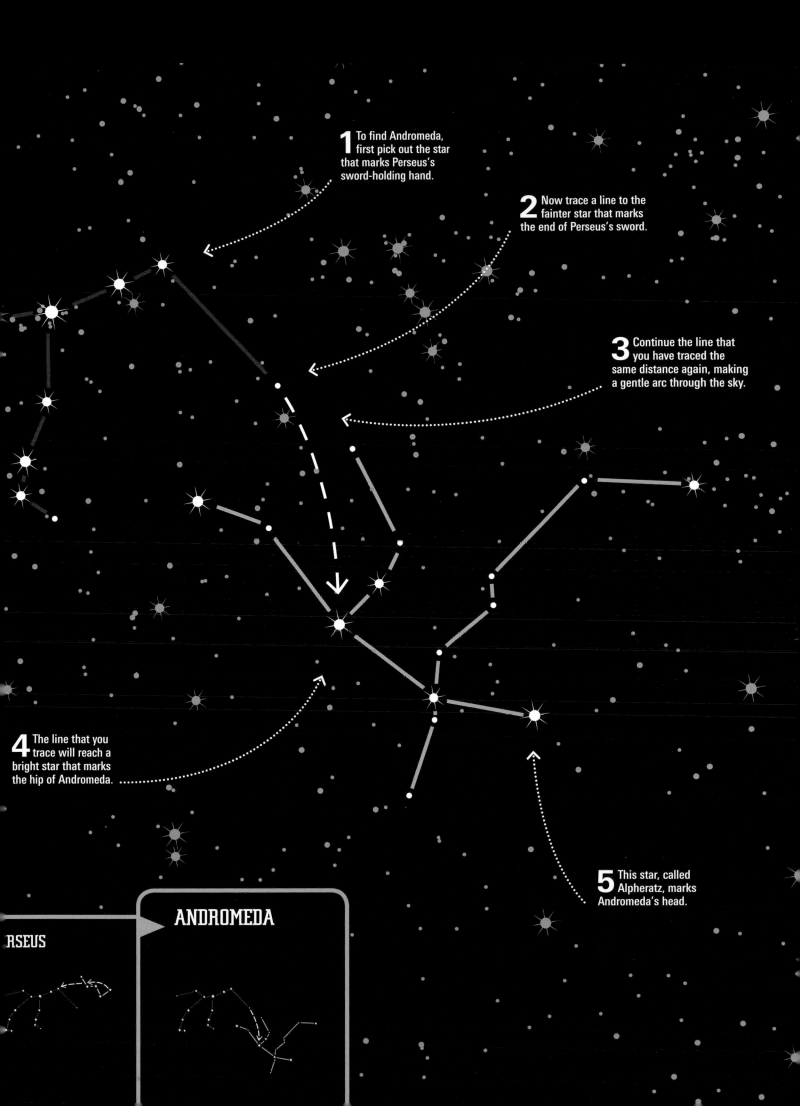

1 To find Andromeda, first pick out the star that marks Perseus's sword-holding hand.

2 Now trace a line to the fainter star that marks the end of Perseus's sword.

3 Continue the line that you have traced the same distance again, making a gentle arc through the sky.

4 The line that you trace will reach a bright star that marks the hip of Andromeda.

5 This star, called Alpheratz, marks Andromeda's head.

ANDROMEDA

RSEUS

THE ANDROMEDA GALAXY IS FOUND BETWEEN
THE OUTSTRETCHED ARM AND LEG OF ANDROMEDA.

ANDROMEDA GALAXY

MESSIER 31

The **Andromeda Galaxy** is a spiral galaxy within the constellation Andromeda. It is the **nearest and brightest spiral galaxy** to our galaxy, the Milky Way, and appears as a smudge in the sky. At 2.5 million light-years away, it is the **most distant object** that can be seen with the naked eye.

▲ This 2008 image of the Andromeda Galaxy (Messier 31), taken using a telescope in France, shows the beautiful natural colours of our nearest spiral galaxy.

Finding the Andromeda Galaxy

The Andromeda Galaxy lies beside the right leg of Andromeda. To the naked eye, it looks like a smudge of light as wide as a full moon. Binoculars or a telescope will reveal more of the spiral galaxy's detail.

ANDROMEDA GALAXY ⋯⋯⋯⋯⋯

THE CONSTELLATION PEGASUS BRANCHES OFF THE STAR THAT MARKS ANDROMEDA'S HEAD.

PEGASUS
THE WINGED HORSE

Pegasus represents a **mythical flying horse**. He was born from the body of the **Gorgon Medusa** when she was killed by Perseus. Pegasus was tamed and ridden by the mythical hero **Bellerophon**.

A hero's horse

Pegasus was ridden by the mythical hero Bellerophon on his mission to slay the Chimaera, a monstrous fire-breathing creature. Full of his success, Bellerophon tried to ride Pegasus to Olympus to join the gods. He fell back to Earth while Pegasus made it to the top.

MORE THAN 30 FULL MOONS WOULD FIT INTO THE GREAT SQUARE OF PEGASUS

Representing the **mythical flying horse**, the stars of Pegasus can be easily picked out in the sky. The constellation is made up of a vast square of stars, named the **Great Square of Pegasus**, and has three branches to make up the horse's legs and head.

GREAT SQUARE OF PEGASUS

PEGASUS

1 Find Pegasus by first looking for the star that marks Andromeda's head. This star belongs to the Andromeda constellation. ....

YOUR ROUTE ACROSS THE SKY

CASSIOPEIA

CEPHEUS

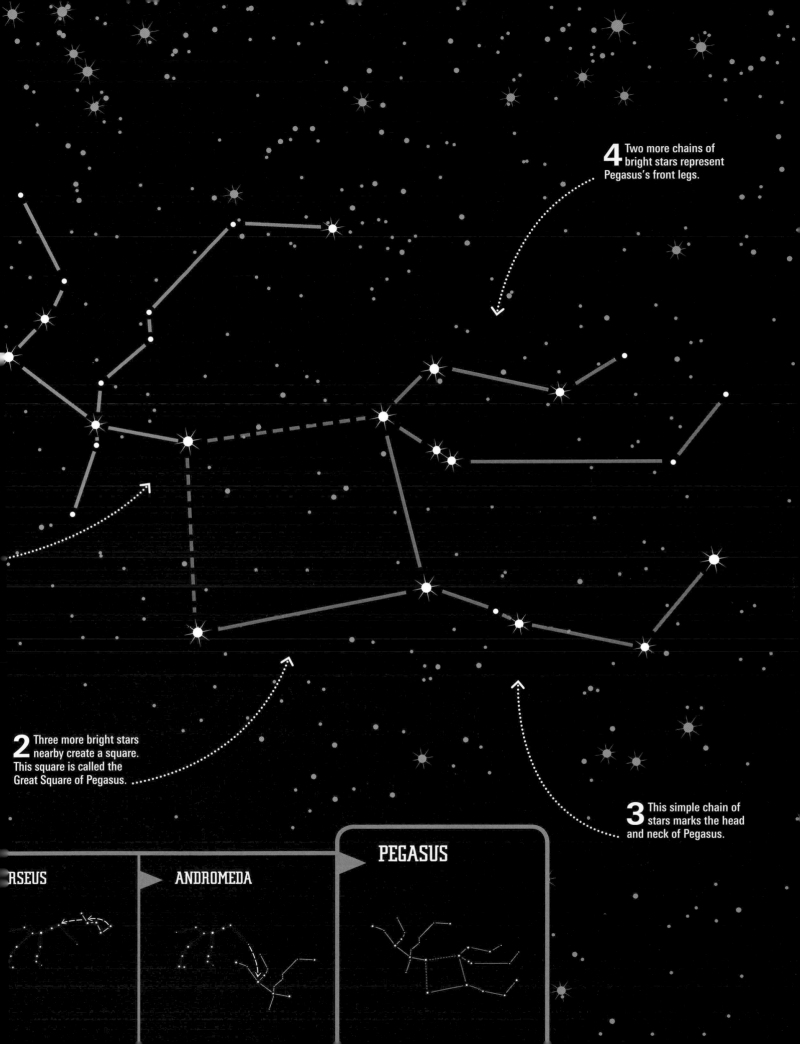

4 Two more chains of bright stars represent Pegasus's front legs.

2 Three more bright stars nearby create a square. This square is called the Great Square of Pegasus.

3 This simple chain of stars marks the head and neck of Pegasus.

RSEUS

ANDROMEDA

PEGASUS

STEPHAN'S QUINTET IS A GROUP OF FIVE GALAXIES
FOUND BELOW THE LEGS OF PEGASUS.

STEPHAN'S QUINTET
COMPACT GALAXY GROUP

The **first compact group of galaxies** discovered, **Stephan's Quintet** is a group of five galaxies. Four of the galaxies lie very near each other in space, while the bluish one at the top of the image is actually much closer to Earth. The two galaxies in the centre are **passing through each other**.

▲ Images taken by the Hubble Space Telescope and Subaru Telescope have been combined to create this image of the visible and infrared light emitted by Stephan's Quintet.

Finding Stephan's Quintet

Stephan's Quintet is located in Pegasus. To find the group, look for the star at the top right of the Great Square of Pegasus, then find the bright star above it to the right. With a telescope you may see a small smudge in the sky, but the five separate galaxies can be seen only through an observatory telescope.

STEPHAN'S QUINTET

FIND ARIES BY TRACING A LINE FROM THE GREAT SQUARE OF PEGASUS.

ARIES
THE RAM

Aries represents a **mythical ram** whose beautiful hide (fleece) was sought by the ancient Greek hero **Jason and the Argonauts**. Aries is one of the 12 zodiac constellations.

The Argonauts
The Argonauts were a group of Greek heroes. Sailing on a ship called the Argo with Jason, they searched for the Golden Fleece – the magnificent hide of a ram that was guarded day and night by a dragon.

ARIES HAS BEEN DEPICTED AS A RAM SINCE ANCIENT TIMES

The brightest stars found in the **constellation Aries** can be linked together to form a **crooked line** in the night sky. It is a faint constellation, so is not always easy to see. One end of the line is the body of the ram, while the other end makes up its **head and horns**.

ARIES

4 This bright star represents the head of Aries the ram.

YOUR ROUTE ACROSS THE SKY

CASSIOPEIA

CEPHEUS

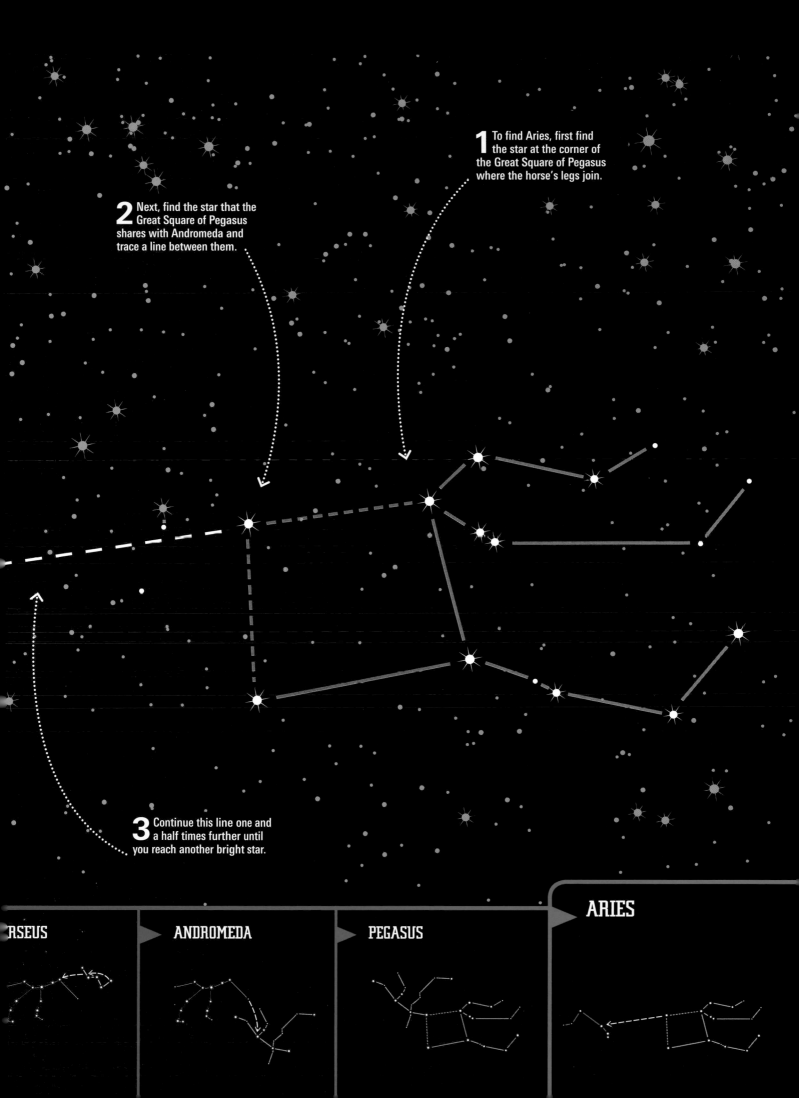

1 To find Aries, first find the star at the corner of the Great Square of Pegasus where the horse's legs join.

2 Next, find the star that the Great Square of Pegasus shares with Andromeda and trace a line between them.

3 Continue this line one and a half times further until you reach another bright star.

RSEUS

ANDROMEDA

PEGASUS

ARIES

AUTUMN IS THE BEST TIME TO SEARCH THE SKIES FOR THE CONSTELLATIONS IN ROUTE THREE.

REVIEW ROUTE THREE
CASSIOPEIA TO ARIES

When you look at all of the constellations of **route three**, you can see that they are closely grouped together. Look for them during **autumn months** when they lie **high in the sky**. You will be able to see some of them at other times of the year too.

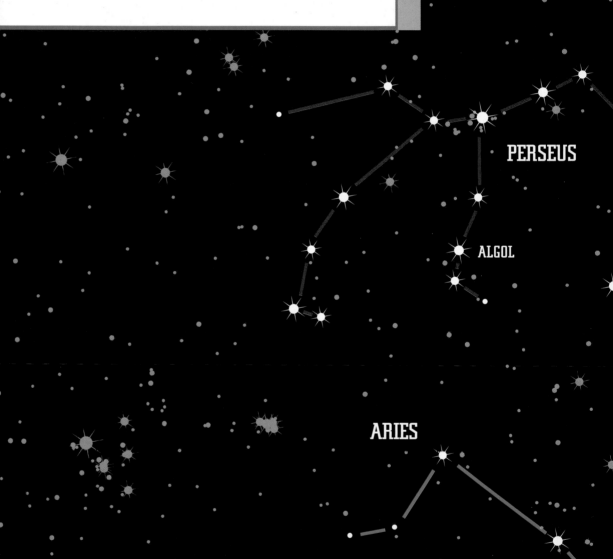

PERSEUS

ALGOL

ARIES

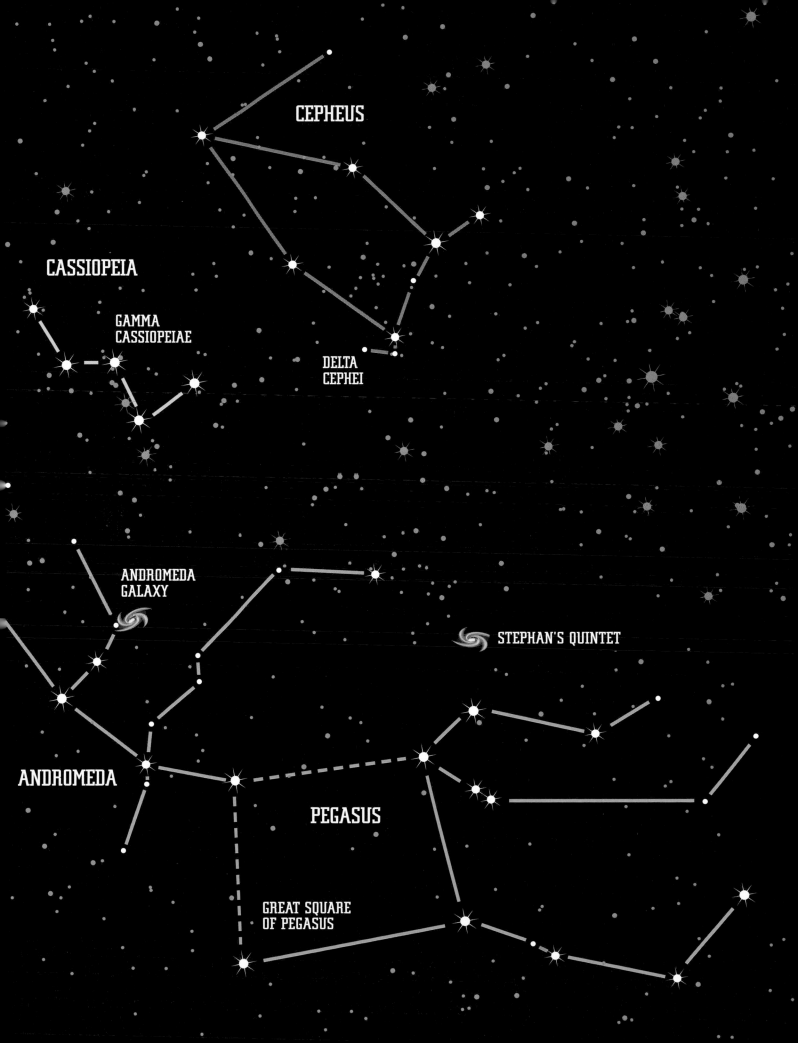

CEPHEUS

CASSIOPEIA

GAMMA
CASSIOPEIAE

DELTA
CEPHEI

ANDROMEDA
GALAXY

STEPHAN'S QUINTET

ANDROMEDA

PEGASUS

GREAT SQUARE
OF PEGASUS

88

SEE IF YOU CAN SPOT THE CONSTELLATIONS OF ROUTE THREE IN THIS VIEW OF THE NIGHT SKY.

FIND THE CONSTELLATIONS
CASSIOPEIA TO ARIES

Try finding the **route three** constellations in this view of the stars, before looking at the **real night sky**. This view of the sky does not include the Plough, so **start by looking for Cassiopeia**.

YOUR ROUTE ACROSS THE SKY

CASSIOPEIA

CEPHEUS

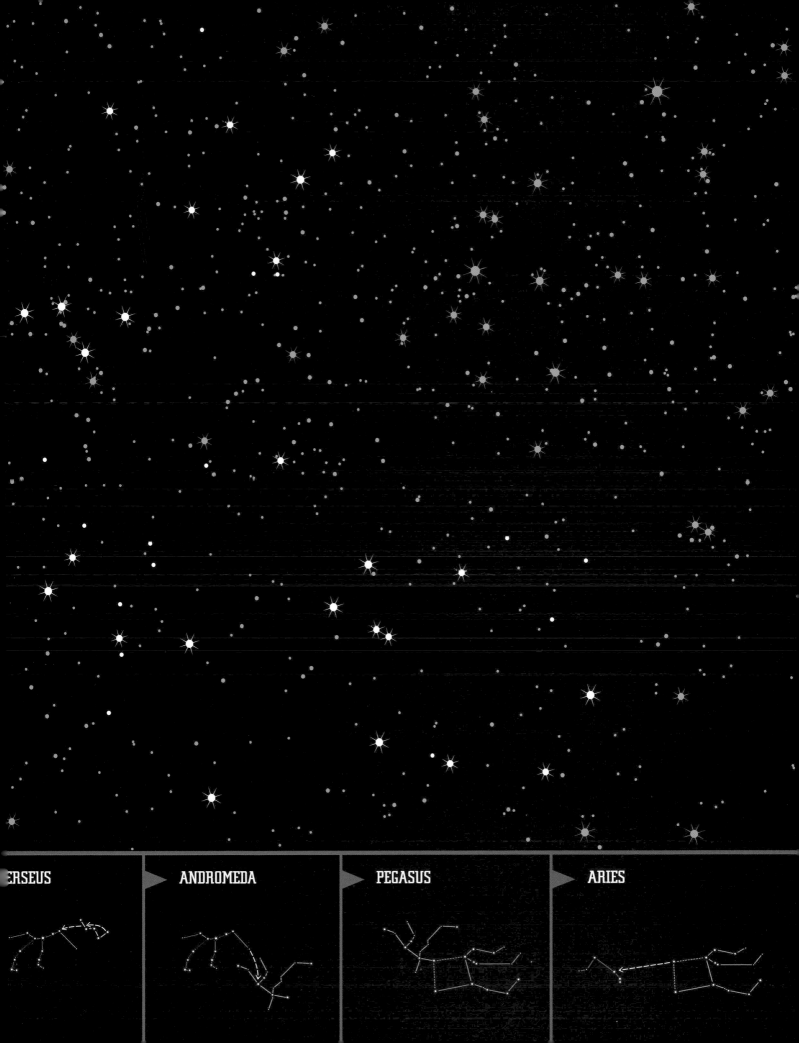

ERSEUS

ANDROMEDA

PEGASUS

ARIES

ROUTE 4

CYGNUS TO SERPENS CAPUT

IN THE SUMMER, STARHOP FROM CYGNUS TO SIX OTHER STAR PATTERNS, INCLUDING THE THREE BRIGHT STARS THAT MAKE UP THE SUMMER TRIANGLE, AND FIND THE STUNNING RING NEBULA.

CYGNUS

LYRA

AQUILA

SUMMER TRIANGLE

OPHIUCHUS

SERPENS CAUDA

SERPENS CAPUT

Aquila
The eagle constellation Aquila can be spotted high in the sky in the centre of this image. Its brightest star, Altair, stands out in the summer night sky.

CYGNUS IS FOUND BY TRACING A LINE FROM THE THREE BRIGHTEST STARS IN CASSIOPEIA.

CYGNUS
THE SWAN

Stretching out along the path of the **Milky Way**, **Cygnus** (pronounced sig-nus), or the **northern cross**, is a cross-shaped constellation depicting a **flying swan** from ancient Greek mythology.

Star colours

The star that marks the head of Cygnus is Albireo, which is actually a double star. Using a telescope, you can see that one of the stars is blue, while the other is golden.

THE FIRST CONFIRMED BLACK HOLE LIES IN CYGNUS

CYGNUS

DENEB

Depicting a **swan in flight**, **Cygnus** is a large **cross-shaped** constellation near Cassiopeia. The constellation's brightest star, **Deneb**, marks the swan's tail, with three stars representing the swan's neck and head. Two more strings of stars lie across the middle of Cygnus to mark the swan's **outstretched wings**, so it seems to fly along the Milky Way.

YOUR ROUTE ACROSS THE SKY

CYGNUS

2 Look for the three brightest stars in Cassiopeia and trace a line between them.

1 Cassiopeia is a helpful starting point for finding Cygnus in the night sky.

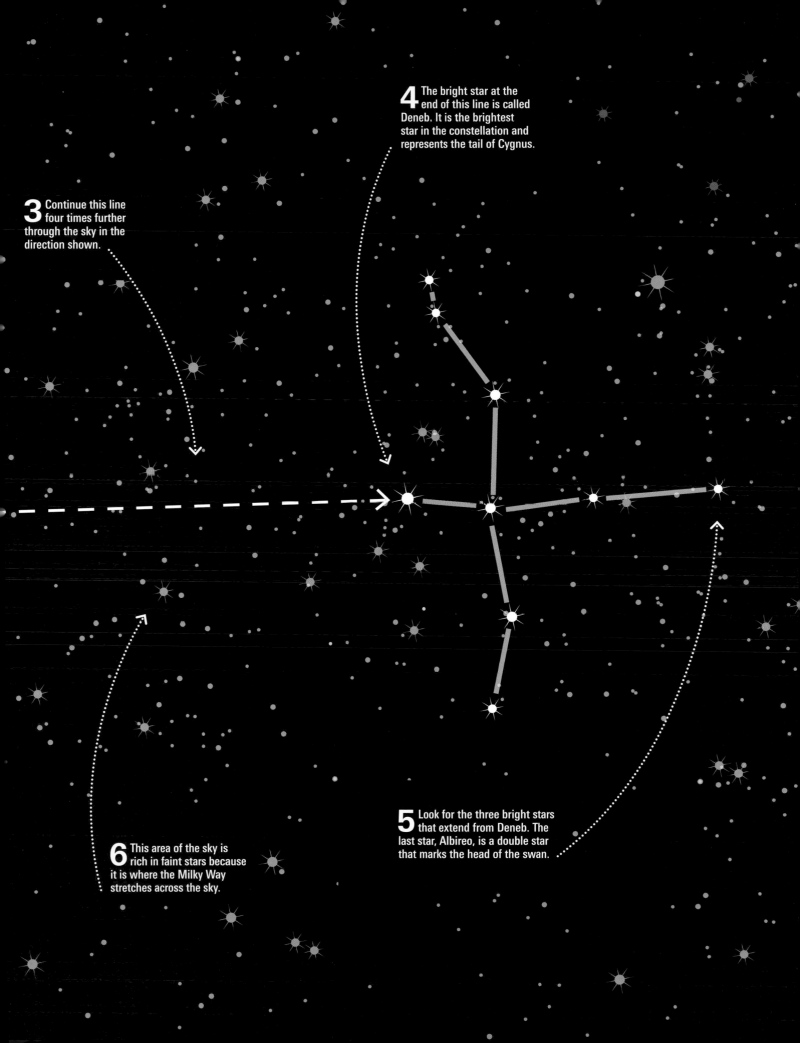

4 The bright star at the end of this line is called Deneb. It is the brightest star in the constellation and represents the tail of Cygnus.

3 Continue this line four times further through the sky in the direction shown.

5 Look for the three bright stars that extend from Deneb. The last star, Albireo, is a double star that marks the head of the swan.

6 This area of the sky is rich in faint stars because it is where the Milky Way stretches across the sky.

94

LYRA CAN BE FOUND JUST ABOVE THE TIP OF CYGNUS'S LEFT WING.

LYRA
THE LYRE

Lyra, or the lyre, is depicted as the musical instrument of the **poet and musician Orpheus** of Greek mythology. The constellation's brightest star, **Vega**, is the **fifth brightest** in the night sky.

Orpheus and Eurydice

Lyra represents the instrument played by Orpheus. He used his lyre to charm Hades, Greek god of the Underworld, so that Orpheus could rescue his wife, Eurydice.

ARAB ASTRONOMERS SAW THE SHAPE OF LYRA AS AN EAGLE OR VULTURE

1 To find Lyra, first locate the star Deneb that marks Cygnus's tail.

Lyra depicts a **stringed musical instrument** called a **lyre**. It is a small and unusual arrangement of stars, most easily identified if you first look for the star **Vega**, the brightest star in the constellation. From Vega, other strings of stars branch out at random to mark the **strings of the lyre**.

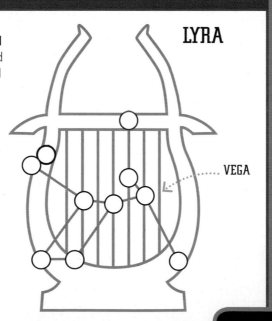

LYRA

VEGA

YOUR ROUTE ACROSS THE SKY

CYGNUS

LYRA

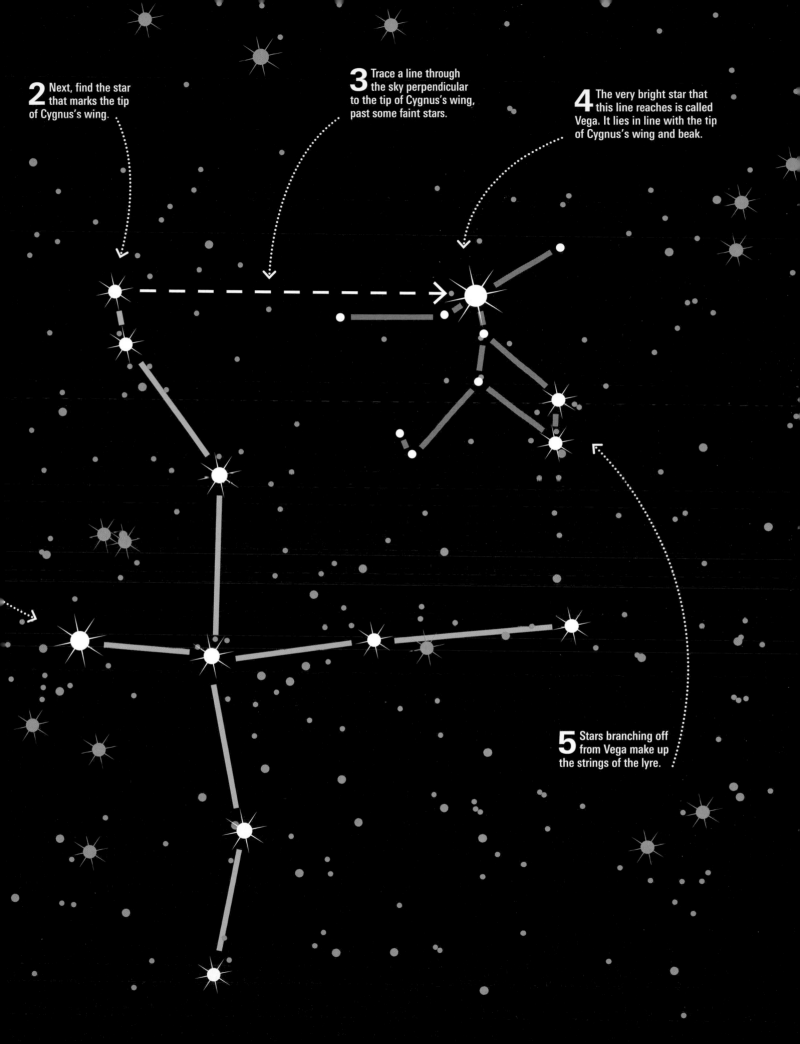

2 Next, find the star that marks the tip of Cygnus's wing.

3 Trace a line through the sky perpendicular to the tip of Cygnus's wing, past some faint stars.

4 The very bright star that this line reaches is called Vega. It lies in line with the tip of Cygnus's wing and beak.

5 Stars branching off from Vega make up the strings of the lyre.

THE RING NEBULA IS FOUND BETWEEN THE TWO
BRIGHT STARS THAT LIE BELOW VEGA IN LYRA.

RING NEBULA
MESSIER 57

The **Ring Nebula** is a **planetary nebula**, formed
4,000 years ago when a **Sun-like star** began to
run out of hydrogen fuel and swelled up into a large,
cool star known as a **red giant**. Its outer layers
were ejected into space, leaving the star's hot
core exposed as a **white dwarf** that continues
to illuminate the beautiful shells of gas around it.

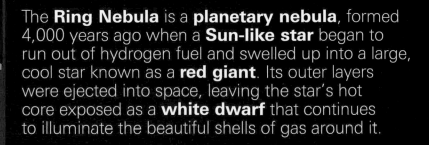

▲ This image of the
Ring Nebula (Messier
57), captured by the
Hubble Space
Telescope, shows the
magnificent colours
of its different gases.

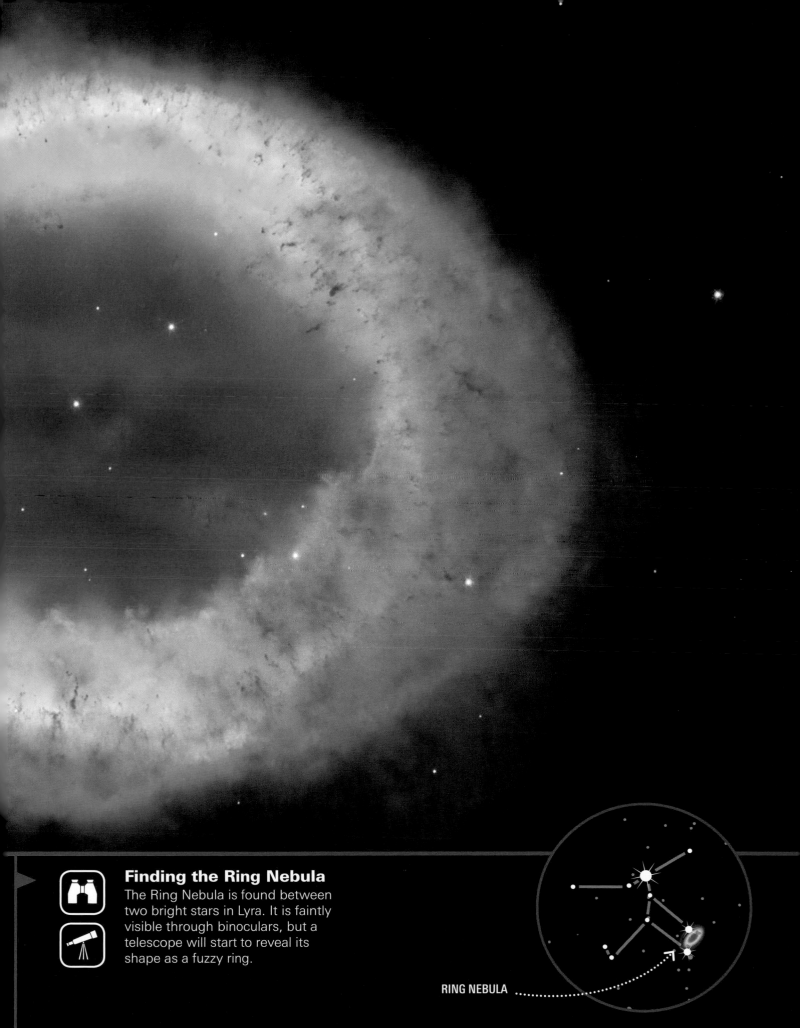

Finding the Ring Nebula
The Ring Nebula is found between two bright stars in Lyra. It is faintly visible through binoculars, but a telescope will start to reveal its shape as a fuzzy ring.

RING NEBULA

THE TIP OF AQUILA'S RIGHT WING CAN BE FOUND BY FOLLOWING THE LINE OF CYGNUS'S NECK.

1 To find Aquila, first look for Deneb, the bright star at Cygnus's tail.

AQUILA THE EAGLE

The constellation **Aquila** is depicted as the **eagle of the Greek god Zeus**. The constellation seems to **soar through the night sky** along the bright path of the Milky Way.

Thunderbird
One ancient Greek myth tells that Aquila was the bird of the god Zeus. Aquila's job was to collect and carry the thunderbolts that Zeus threw at his enemies.

THE BRIGHT STAR ALTAIR FORMS ONE CORNER OF THE SUMMER TRIANGLE

Aquila is a simple constellation that is depicted as a **flying eagle**. Its brightest star, **Altair**, marks the base of the eagle's neck. The other stars in the constellation can be linked together to form the **shape of Aquila's two wings**, with a branch that represents the tail.

ALTAIR

AQUILA

YOUR ROUTE ACROSS THE SKY

CYGNUS

LYRA

AQUILA

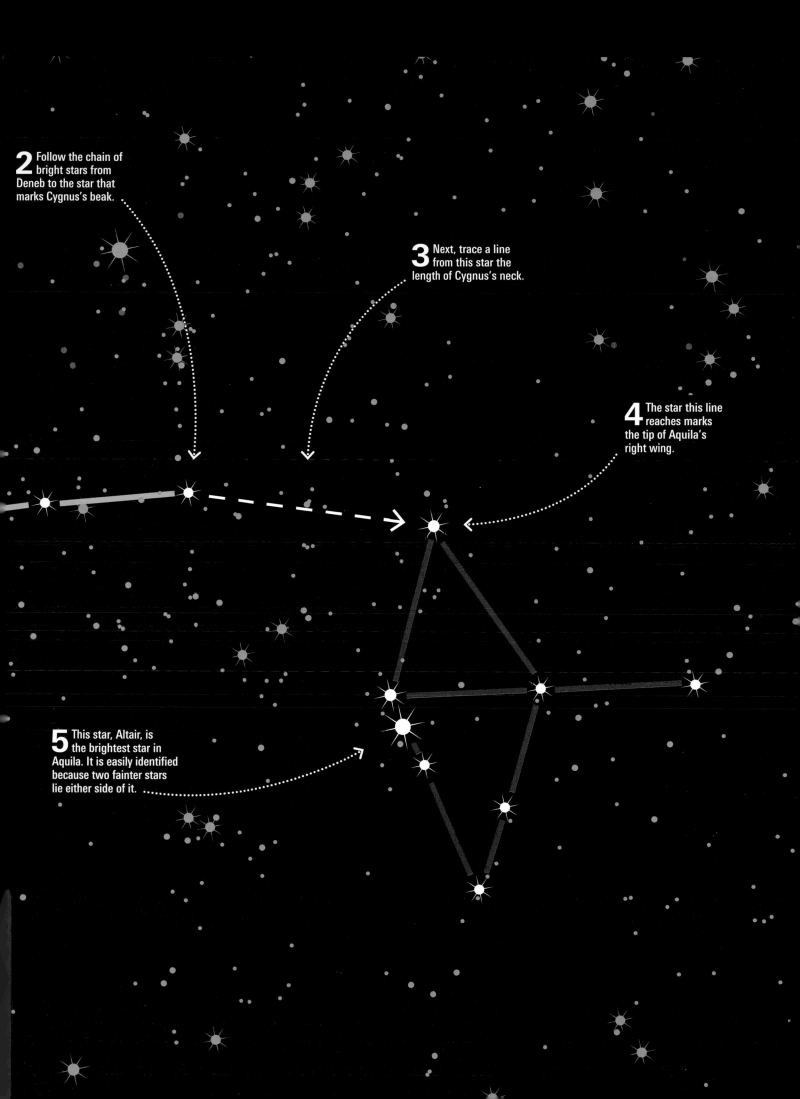

2 Follow the chain of bright stars from Deneb to the star that marks Cygnus's beak.

3 Next, trace a line from this star the length of Cygnus's neck.

4 The star this line reaches marks the tip of Aquila's right wing.

5 This star, Altair, is the brightest star in Aquila. It is easily identified because two fainter stars lie either side of it.

100

THE SUMMER TRIANGLE IS MADE FROM THE BRIGHTEST STARS IN CYGNUS, LYRA, AND AQUILA.

SUMMER TRIANGLE
A SUMMER ASTERISM

The Summer Triangle is an asterism that can be seen in the northern night sky. It is a simple triangle formed from **three of the brightest stars** in the **summer** night sky – Deneb, Vega, and Altair.

Twinkling stars
Stars in the night sky appear to twinkle when we view them from Earth because the light that they emit gets disturbed as it enters Earth's thick atmosphere.

CYGNUS'S STAR DENEB IS 60,000 TIMES MORE LUMINOUS THAN THE SUN

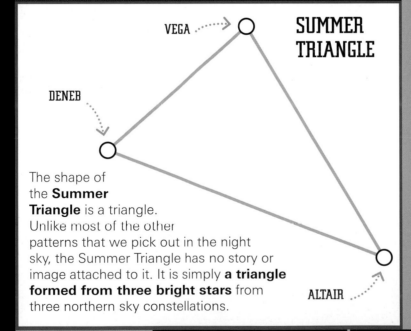

VEGA

DENEB

SUMMER TRIANGLE

ALTAIR

The shape of the **Summer Triangle** is a triangle. Unlike most of the other patterns that we pick out in the night sky, the Summer Triangle has no story or image attached to it. It is simply **a triangle formed from three bright stars** from three northern sky constellations.

1 To identify the Summer Triangle, first find the bright star Deneb at the tail of Cygnus. This is the first point of the Summer Triangle.

YOUR ROUTE ACROSS THE SKY

CYGNUS

LYRA

AQUILA

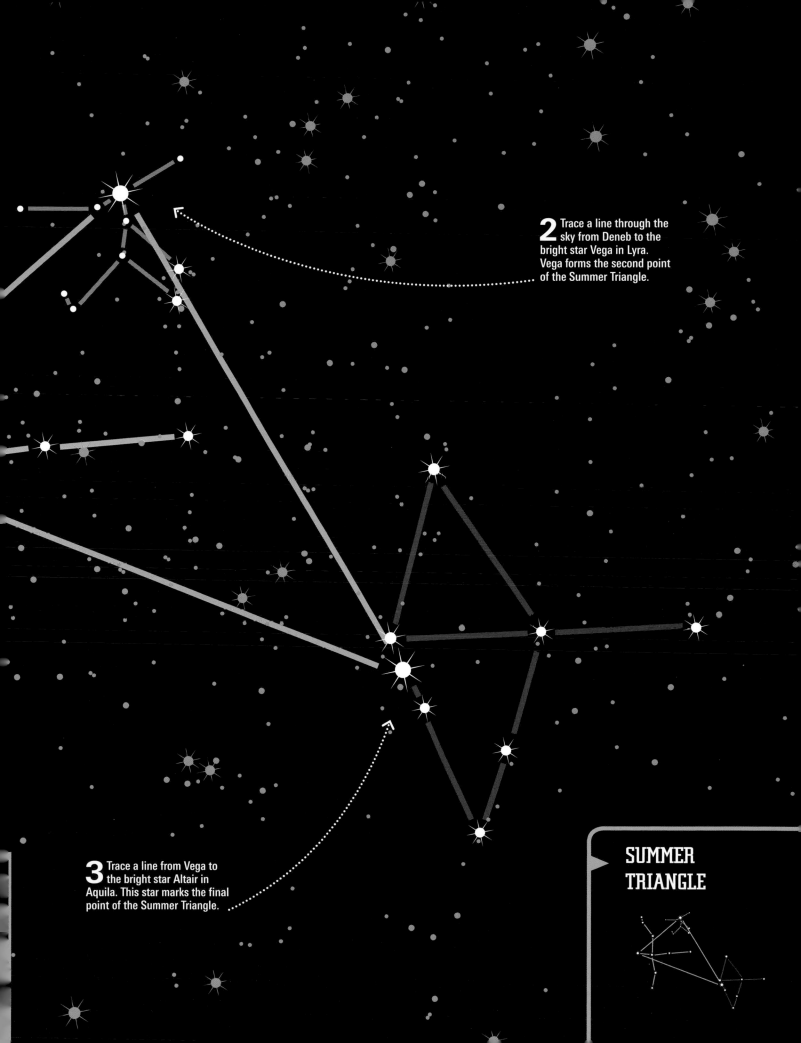

2 Trace a line through the sky from Deneb to the bright star Vega in Lyra. Vega forms the second point of the Summer Triangle.

3 Trace a line from Vega to the bright star Altair in Aquila. This star marks the final point of the Summer Triangle.

SUMMER TRIANGLE

OPHIUCHUS IS FOUND BY FOLLOWING THE WING OF AQUILA TO THE STAR THAT MARKS OPHIUCHUS'S HEAD.

OPHIUCHUS
THE SERPENT HOLDER

Ophiuchus (pronounced off-ee-you-cus) is a large constellation that depicts **Asclepius**, a **mythical healer** who was said to have the power to **revive the dead**. Asclepius is shown **holding a serpent**, a traditional symbol of healing, which is represented by the constellation Serpens.

Kepler's Star

Ophiuchus is the site of the most recent star explosion to occur in the Milky Way. When it exploded in 1604, this previously faint star outshone every other star in the sky and stayed visible for more than a year. The star was named after the man who observed the supernova explosion, Johannes Kepler.

OPHIUCHUS

The stars of **Ophiuchus** can be joined together to form the shape of a **man holding a snake**. A ring of stars marks the body of Ophiuchus, with two branches coming off the ring that represent one of his legs and one of his arms.

3 Continue this line for double the distance beyond Aquila's wing.

2 Next, trace a line from Altair to the star that marks the tip of Aquila's right wing.

1 To find Ophiuchus, first find the bright star Altair in the constellation Aquila.

EVEN THOUGH THE SUN PASSES THROUGH OPHIUCHUS, IT IS NOT AN ASTROLOGICAL ZODIAC SIGN

YOUR ROUTE ACROSS THE SKY

CYGNUS

LYRA

AQUILA

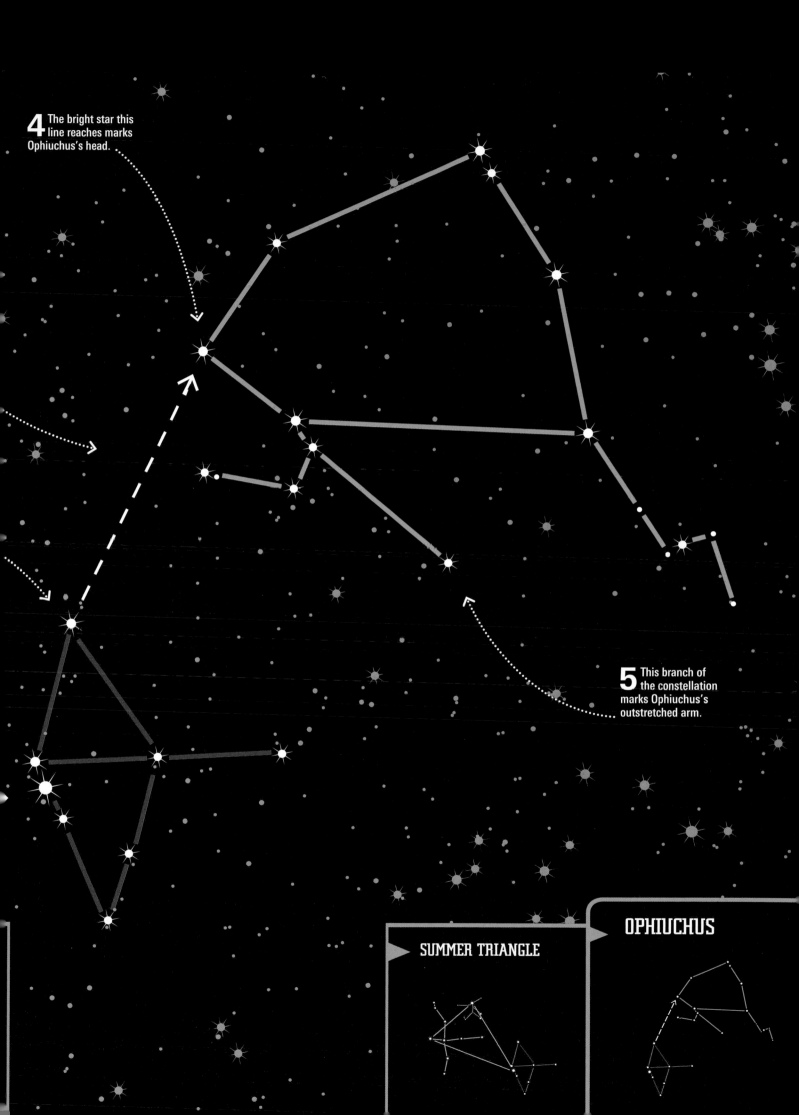

4 The bright star this line reaches marks Ophiuchus's head.

5 This branch of the constellation marks Ophiuchus's outstretched arm.

SUMMER TRIANGLE

OPHIUCHUS

104

THE THREE STARS OF SERPENS CAUDA ARE FOUND EITHER SIDE OF OPHIUCHUS'S OUTSTRETCHED HAND.

1 To find Serpens Cauda, first look for the bright star that marks Ophiuchus's head.

SERPENS CAUDA
THE SERPENT'S TAIL

Serpens Cauda is one half of the constellation **Serpens**. Unlike any other constellation, Serpens is **split into two separate areas**. Serpens Cauda is formed of three stars and depicts the **tail of the huge snake** held by the constellation Ophiuchus.

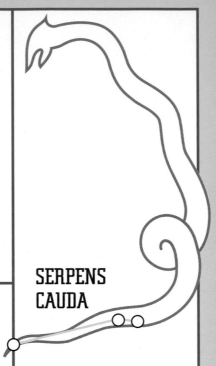

SERPENS CAUDA

Divided stars
Even though it appears to be a constellation of its own, Serpens Cauda is actually just one half of a larger constellation called Serpens. The snake is split so that its tail lies on one side of the constellation Ophiuchus, while its head, known as Serpens Caput, lies on the other.

Serpens Cauda is made up of **three stars**, which are joined together to depict the **tail of a snake**. The rest of the snake's body lies in the other half of the constellation, known as **Serpens Caput**.

2 Trace a line through the four bright stars that mark Ophiuchus's head, shoulder, and arm to reach his hand.

SNAKES ARE A SYMBOL OF REBIRTH BECAUSE THEY SHED THEIR SKINS

YOUR ROUTE ACROSS THE SKY

CYGNUS

LYRA

AQUILA

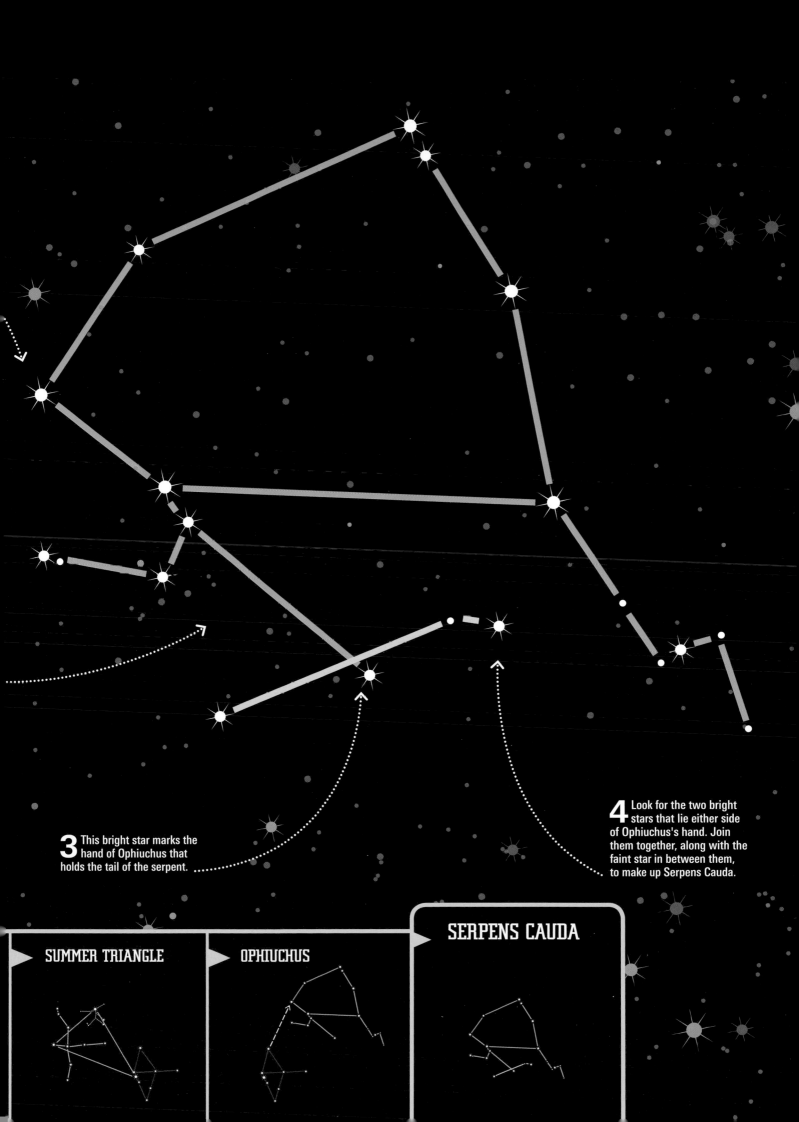

3 This bright star marks the hand of Ophiuchus that holds the tail of the serpent.

4 Look for the two bright stars that lie either side of Ophiuchus's hand. Join them together, along with the faint star in between them, to make up Serpens Cauda.

SUMMER TRIANGLE

OPHIUCHUS

SERPENS CAUDA

106

FIND SERPENS CAPUT BY TRACING A LINE BEYOND THE BASE OF OPHIUCHUS TO THE NECK OF THE SNAKE.

5 This group of three bright stars makes up the head of the serpent. ⋯⋯⋯⋯

SERPENS CAPUT
THE SERPENT'S HEAD

Serpens Caput is the **larger** of the two asterisms that make up the constellation **Serpens**. Lying on the other side of Ophiuchus to Serpens Cauda, Serpens Caput depicts the **head of the huge snake** that is held by Asclepius in the constellation Ophiuchus.

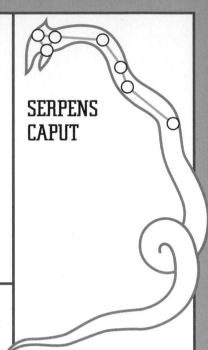

SERPENS CAPUT

Healing the dead

Ancient Greek myths suggest that Asclepius learnt how to revive the dead by watching two snakes. Having killed a snake, Asclepius watched as another snake placed a herb on it, restoring it to full health. Asclepius tried the same technique on people and discovered he could heal the dead in this way too.

Serpens Caput is a simple chain of seven stars that depict the **head of a snake**. The other half of the snake's body lies in Serpens Cauda, on the other side of the constellation Ophiuchus.

SERPENS IS THE ONLY CONSTELLATION TO BE SPLIT INTO TWO PARTS

YOUR ROUTE ACROSS THE SKY

▶▶▶

CYGNUS

LYRA

AQUILA

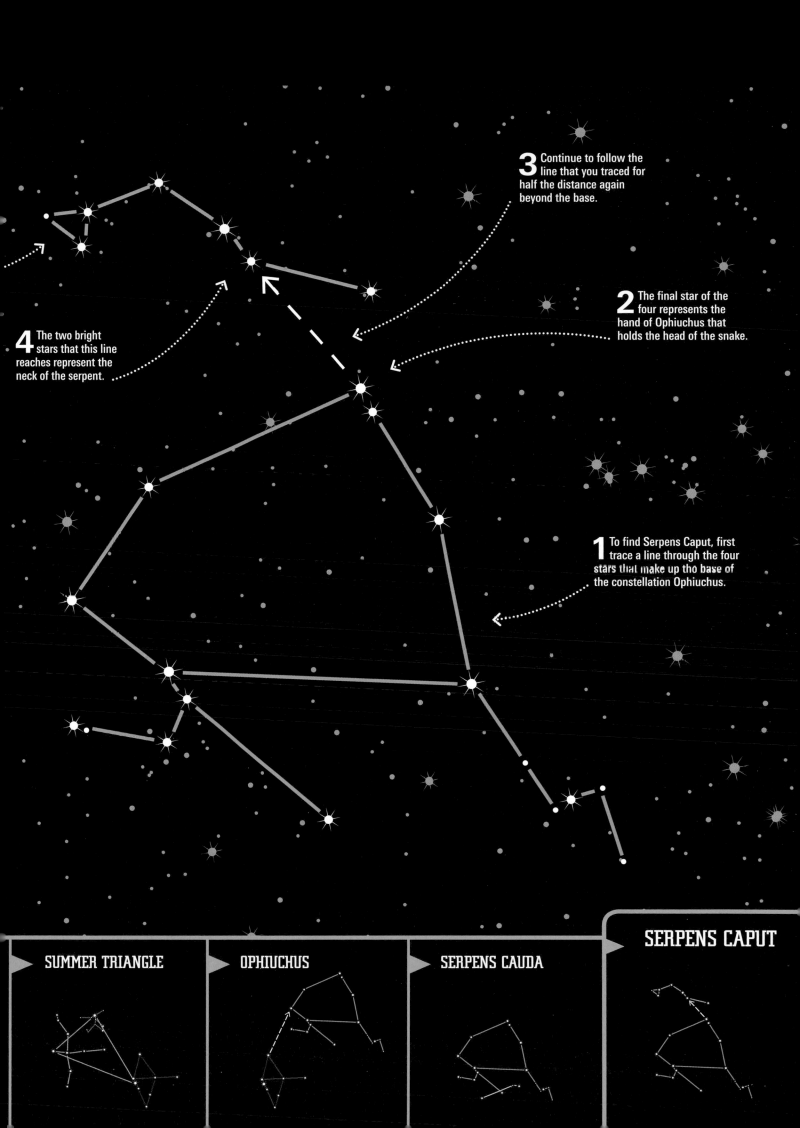

3 Continue to follow the line that you traced for half the distance again beyond the base.

2 The final star of the four represents the hand of Ophiuchus that holds the head of the snake.

4 The two bright stars that this line reaches represent the neck of the serpent.

1 To find Serpens Caput, first trace a line through the four stars that make up the base of the constellation Ophiuchus.

SUMMER TRIANGLE

OPHIUCHUS

SERPENS CAUDA

SERPENS CAPUT

THE GLOBULAR STAR CLUSTER MESSIER 5 IS FOUND TO THE SIDE OF THE NECK OF SERPENS CAPUT.

MESSIER 5
GLOBULAR STAR CLUSTER

Messier 5 is a **globular star cluster**, a tightly packed ball of **hundreds of thousands of stars** that lies in the halo of our galaxy, the Milky Way. Along with other globular clusters, Messier 5 is one of the **oldest objects** in our galaxy, at about **13 billion years old**.

▲ The Hubble Space Telescope took this image of Messier 5, which lies nearly 25,000 light-years away from Earth.

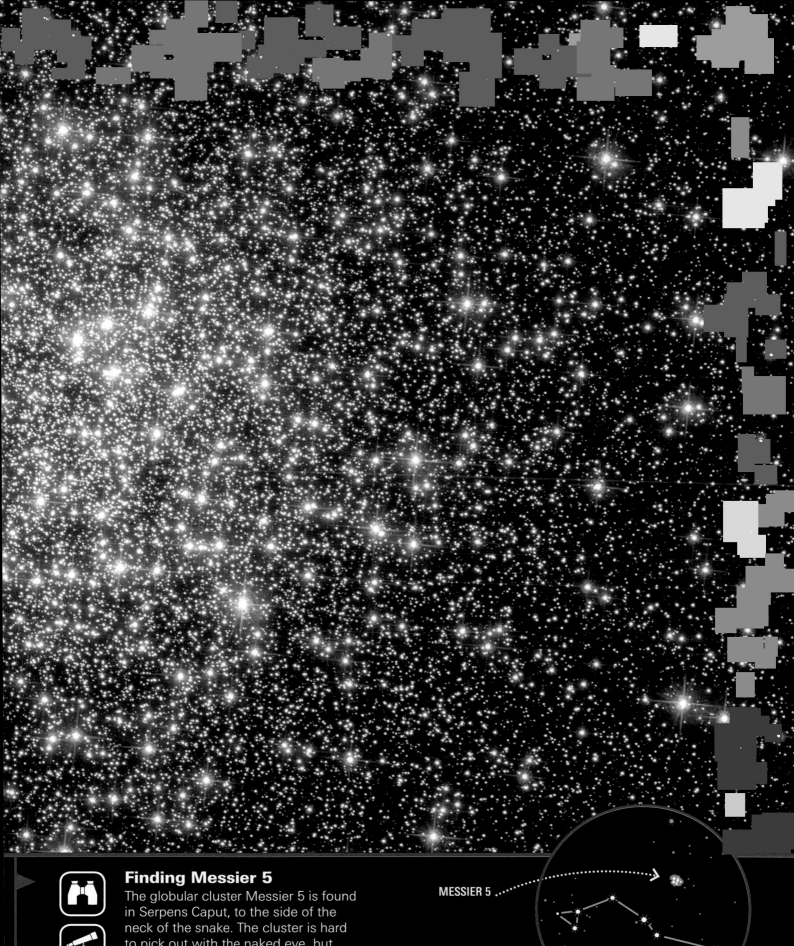

Finding Messier 5
The globular cluster Messier 5 is found in Serpens Caput, to the side of the neck of the snake. The cluster is hard to pick out with the naked eye, but binoculars will show a faint smudge where Messier 5 lies.

MESSIER 5

THE CONSTELLATIONS OF ROUTE FOUR CAN BE SEEN IN SUMMER SKIES.

REVIEW ROUTE FOUR
CYGNUS TO SERPENS CAPUT

Here is a view of the night sky showing where the **constellations of route four** lie. You can use the route we have learnt to find these constellations during **summer evenings** when they will be high up in the sky.

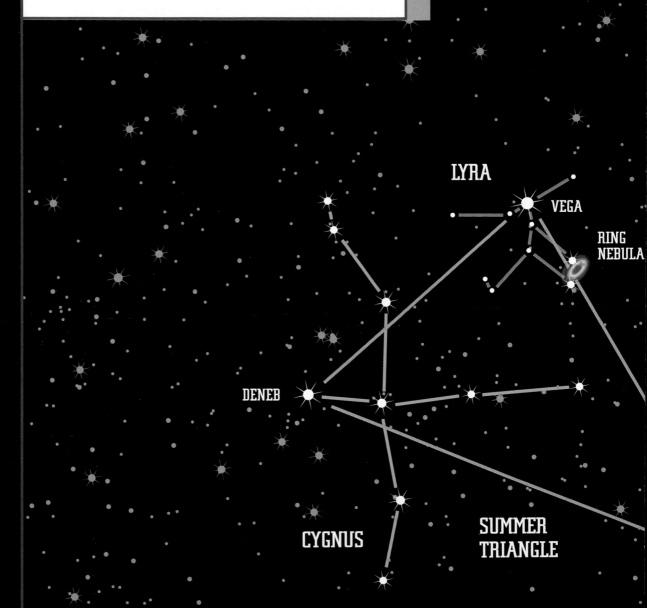

LYRA

VEGA

RING NEBULA

DENEB

CYGNUS

SUMMER TRIANGLE

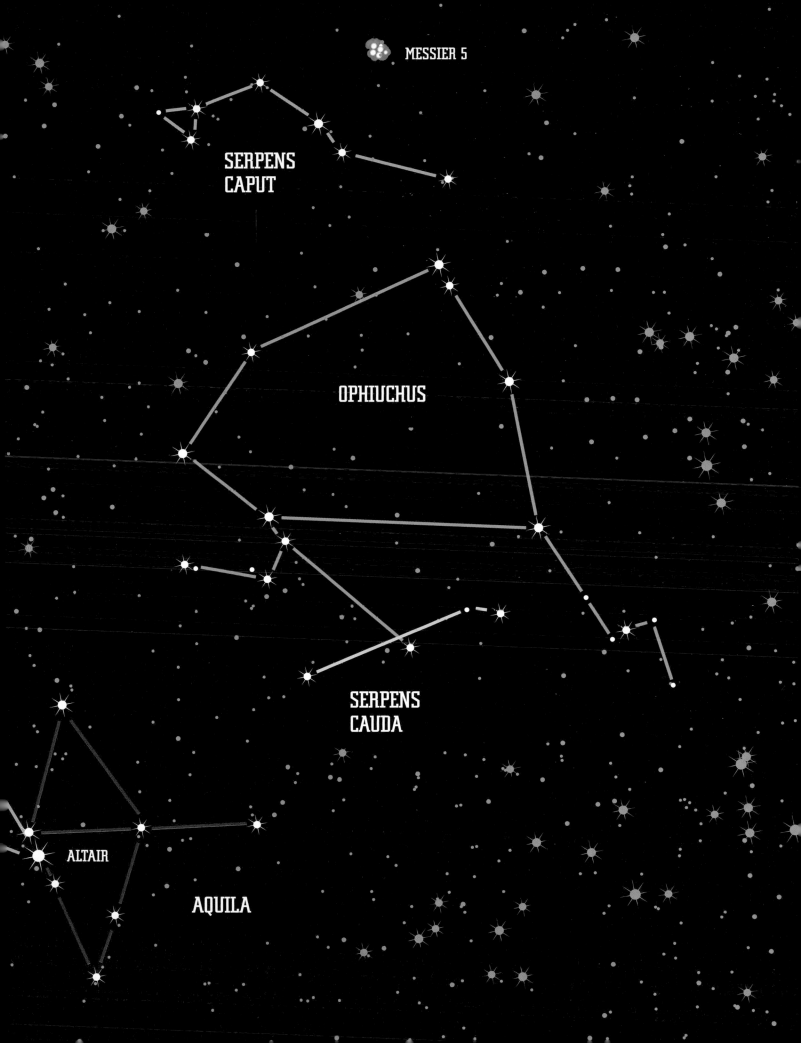

MESSIER 5

SERPENS
CAPUT

OPHIUCHUS

SERPENS
CAUDA

ALTAIR

AQUILA

USE THIS VIEW OF THE NIGHT SKY TO PRACTISE
THE PATH YOU HAVE LEARNT FOR ROUTE FOUR.

FIND THE CONSTELLATIONS

CYGNUS TO SERPENS CAPUT

The path for route four along the bottom
of these pages will guide you through this
view of the sky. This view of the sky does not
show Cassiopeia. **Practise stargazing** here
before heading outside to try it for yourself.

YOUR
ROUTE
ACROSS
THE SKY

CYGNUS

LYRA

AQUILA

SUMMER TRIANGLE

OPHIUCHUS

SERPENS CAUDA

SERPENS CAPUT

AMONG THE STARS

STARS ARE NOT THE ONLY OBJECTS THAT LIGHT UP THE NIGHT SKY. AS WELL AS THE CONSTELLATIONS, ASTRONOMERS VIEW PLANETS, GALAXIES, AND MANY OTHER DEEP-SKY OBJECTS IN ORDER TO DEVELOP A DEEPER UNDERSTANDING OF OUR UNIVERSE.

Evening star
This image reveals the Moon's stunning surface detail. To its left lies the planet Venus, known as the evening star, and Mercury can be seen below.

The Moon is the **largest object in the night sky**. While it appears to be bright, the Moon emits no light. Instead, it **reflects the light of the Sun**.

THE MOON
LUNAR PHASES AND SOLAR ECLIPSES

MOVEMENT OF THE MOON

The Moon orbits Earth over a period of 27.3 days. As it does so, sections of its face are lit up by the light of the Sun, making it visible in the night sky.

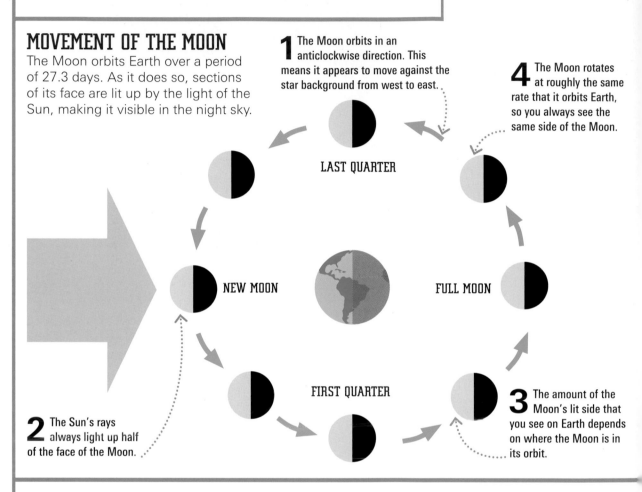

1 The Moon orbits in an anticlockwise direction. This means it appears to move against the star background from west to east.

4 The Moon rotates at roughly the same rate that it orbits Earth, so you always see the same side of the Moon.

LAST QUARTER

NEW MOON

FULL MOON

FIRST QUARTER

2 The Sun's rays always light up half of the face of the Moon.

3 The amount of the Moon's lit side that you see on Earth depends on where the Moon is in its orbit.

PHASES OF THE MOON

As it orbits Earth, the shape of the Moon appears to change. These different shapes, called lunar phases, occur because each day the Moon is in a different position relative to the Sun. The full cycle takes 29.5 days.

1 When the Moon is on the opposite side of Earth to the Sun, its face is fully lit.

2 The Moon is said to be "waning" when it appears to be shrinking.

Full moon   Waning gibbous   Last quarter   Waning cresce

SOLAR ECLIPSE

A solar eclipse occurs when the Sun, the Moon, and Earth are directly aligned so that the Moon blocks sunlight from reaching Earth. A shadow is cast on Earth by the Moon, plunging that part of Earth into darkness for several minutes.

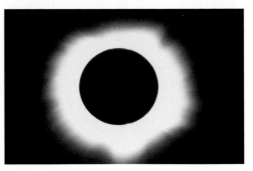

▲ The stage of a solar eclipse when the Sun is completely blocked by the Moon is called totality. This photograph reveals the Sun's outer atmosphere, the corona, during totality.

1 A solar eclipse occurs when the Moon lies directly between the Sun and Earth, blocking the Sun's rays.

2 A shadow is cast on Earth by the Moon. Anyone within the umbra, this darker area of the shadow, will see a total eclipse.

3 Viewers within the lighter area of this shadow, called the penumbra, will see a partial eclipse, as some of the Sun's rays reach Earth.

4 The Sun is much too bright to view with the naked eye, binoculars, or a telescope, even during an eclipse. Never look directly at the Sun, as its glare can permanently damage eyesight.

3 When the Moon lies between Earth and the Sun, the side that faces Earth is not lit by the Sun.

4 The Moon is said to be "waxing" when it appears to be growing.

5 Only half of the Moon is visible when it lies at a right-angle to the Sun.

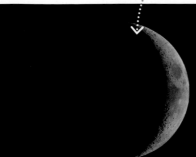

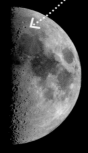

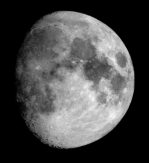

| New moon | Waxing crescent | First quarter | Waxing gibbous |

THE MILKY WAY

VIEWING OUR GALAXY

Looking into the sky on a clear, dark night, you may be able to see a **beautiful milky glow lighting up the night sky**. This glow that stretches across the sky is made by the stars of **the Milky Way**, the huge disc-shaped **spiral galaxy** in which our Solar System lies.

▲ In this photograph of the Milky Way, dark clouds of dust and gas within the galaxy can be seen blocking the light from the stars that shine behind.

YOUR VIEW

Earth lies within one of the arms of the Milky Way, about two-thirds out from the galaxy's centre. When we see the Milky Way in the night sky, we are looking at the edge of the Milky Way's disc.

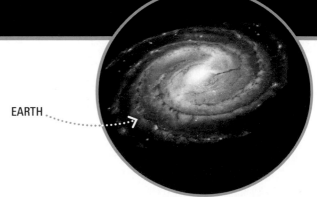

EARTH

PLANET SPOTTING
NIGHT SKY WANDERERS

Ancient astronomers noticed several bright **star-like lights** that moved gradually through the sky against the background of stars. They named these lights "**planets**", meaning "**wanderers**". All of the planets in our Solar System travel along roughly the **same path**, which runs through the **twelve constellations of the zodiac**. Most of the planets can be seen with the naked eye.

NAKED EYE PLANETS

Because of their distance from Earth, we cannot see all of the planets in the Solar System with the naked eye, but we are able to spot Mercury, Venus, Mars, Jupiter, and Saturn (shown right, not to scale). The two most distant planets in the Solar System, Uranus and Neptune, can be seen with a telescope.

WHERE TO FIND THE PLANETS

1 This imaginary line, called the ecliptic, roughly traces the path of the Sun and the planets through our sky. The planets, including Earth, all orbit the Sun on much the same plane, so they all cross our sky along the same path.

2 The ecliptic runs through the constellations of the zodiac – Aries, Taurus, Gemini, Cancer, Leo, Virgo, Libra, Scorpius, Sagittarius, Capricornus, Aquarius, and Pisces. So, the planets will always be found moving through one of these constellations.

3 If you see something that looks like a very bright star along this line that doesn't belong in a constellation, you are probably looking at a planet.

4 There are many helpful websites and mobile phone apps that list when planets will be crossing our skies and which constellation they will be moving through.

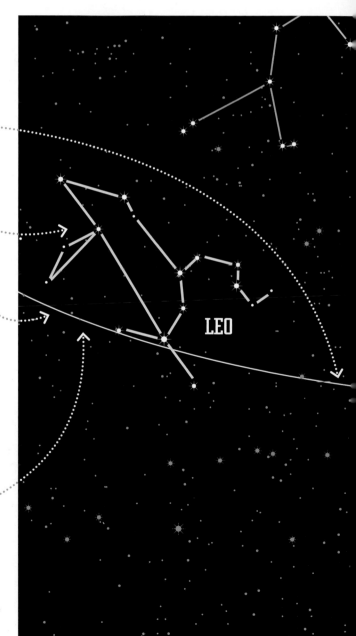

LEO

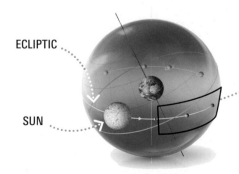

ECLIPTIC

SUN

▲ The ecliptic, the yellow line in this image, traces the path of the Sun and the planets in our Solar System through our sky.

Mercury
Mercury is very difficult to spot, because it is always low in the sky and close to the Sun. It is best observed just before sunset or just after sunrise.

Venus
Venus is an easy planet to spot. Known as the brilliant evening or morning "star", it is the brightest object in the night sky after the Moon. The best time to look for Venus is just before sunset or just after sunrise.

Mars
For much of the time, Mars appears like a reddish star. Every two years and two months, however, there is a two-month window where it is the second-brightest planet in the sky, after Venus.

Jupiter
Jupiter appears brighter than the brightest star in our night sky, Sirius. Using a pair of binoculars, you can even see four of Jupiter's moons, which look like faint stars on either side of it.

Saturn
Saturn looks like a creamy-coloured star and moves very slowly through the sky. Through a telescope, you will be able to see its rings.

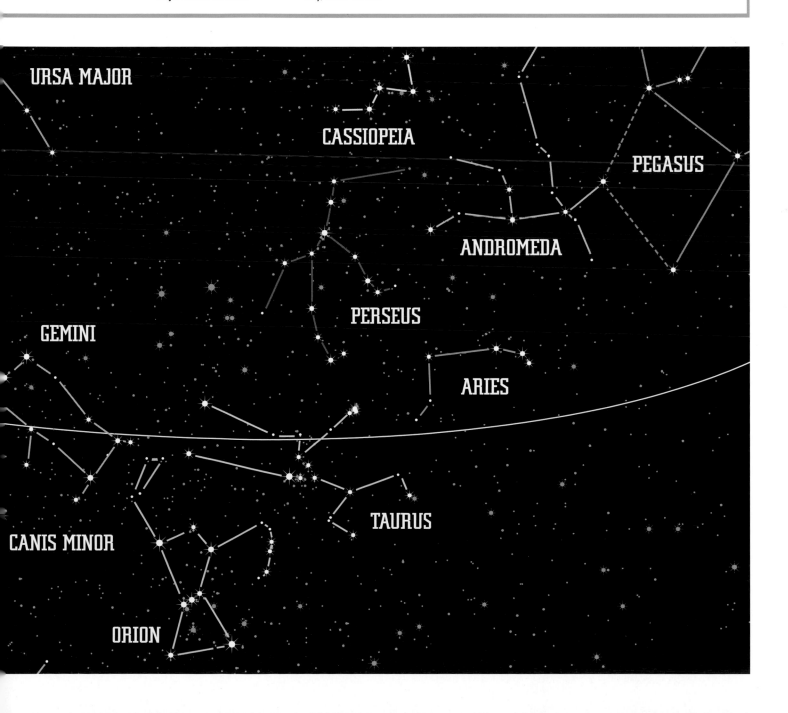

URSA MAJOR

CASSIOPEIA

PEGASUS

ANDROMEDA

PERSEUS

GEMINI

ARIES

CANIS MINOR

TAURUS

ORION

SHOOTING STARS
METEOR SHOWERS

When **specks of dust** enter Earth's atmosphere, they burn up to create **stunning streaks of light** called **shooting stars or meteors**. They often last less than a second and are best seen between midnight and dawn, when Earth is facing away from the Sun and out into space.

▲ The Perseid meteor shower (above) occurs every year in mid August. As many as 80 meteors per hour appear to fly out of the constellation Perseus.

METEOR SHOWERS

Shooting stars can occur any night of the year, but at certain times of year they are more common. During these so-called "meteor showers", many specks of dust burn up in the atmosphere each hour.

ANNUAL METEOR SHOWERS		
Name	**Date**	**Constellation**
Quadrantids	1–6 January	Boötes
Lyrids	19–24 April	Lyra
Eta Aquarids	1–8 May	Aquarius
Delta Aquarids	15 July–15 August	Aquarius
Perseids	25 July–18 August	Perseus
Orionids	16–27 October	Orion
Taurids	20 October–30 November	Taurus
Leonids	15–20 November	Leo
Geminids	7–15 December	Gemini

asterism A pattern formed by stars that are part of one or more constellations.

astronomy A branch of science that studies objects in space, including stars, planets, and galaxies.

celestial Relating to the sky or outer space. An object outside Earth's atmosphere is a celestial body.

celestial equator A circle around the centre of the celestial sphere midway between the two poles. It divides the celestial sphere into two equal halves, one half north of the equator and the other half south.

celestial pole Either of the two points on the celestial sphere directly above Earth's north and south poles. The line joining the celestial poles forms the axis round which the celestial sphere rotates.

celestial sphere An imaginary sphere surrounding Earth, upon which astronomical objects appear to lie.

Cepheid variable star A type of star that regularly changes in brightness. Cepheids are named after Delta Cephei, the first of these stars to be discovered.

comet A small body made up of dust and ice that orbits the Sun. As it gets near the Sun, the ice vaporizes, giving it the appearance of having a glowing head and tail.

constellation An area of sky within boundaries laid down by the International Astronomical Union. There are currently 88 constellations altogether.

deep-sky object A celestial object outside of the Solar System, such as a nebula, star cluster, or galaxy, but not individual stars.

diffuse nebula A cloud of gas and dust illuminated by the stars that lie within it. Most nebulas are diffuse.

double star A pair of stars that appear close together when viewed from Earth. They may have no relation to each other or they may be linked by gravity, making them a binary star.

ecliptic The path along which the Sun appears to travel around the celestial sphere when viewed from Earth.

eclipse An alignment of a planet or moon with the Sun, which casts a shadow on another celestial body. During a solar eclipse, the Moon's shadow is cast on Earth. During a lunar eclipse, Earth's shadow is cast on the Moon.

emission nebula A cloud of gas and dust containing new stars. These give off radiation that makes the gas around them emit light.

galaxy A collection of stars, gas, and dust held together by gravity. Galaxies come in three main shapes: elliptical, spiral, and irregular.

globular star cluster A very tightly packed cluster of old stars.

gravity The force of attraction that pulls all objects that have mass towards one another.

horizon The line in the distance at which Earth's surface and the sky appear to meet.

infrared A type of radiation with a longer wavelength than visible light. Many telescopes use infrared light as it allows them to see different objects in space.

light-year The distance a beam of light can travel in one calendar year: 9,460 billion km (5,878 billion miles).

luminosity (see magnitude)

magnitude The brightness of a celestial object, which can be measured in two ways. An object's apparent magnitude is how bright it appears in the night sky when viewed from Earth. Its absolute magnitude, or luminosity, is the amount of light energy emitted by the object.

Messier object One of more than 100 deep-sky objects catalogued by French astronomer Charles Messier in 1781. He listed these celestial objects so he would not mistake them for comets.

meteor A streak of light, also called a shooting star, caused by a small chunk of space rock or dust burning up in Earth's atmosphere.

meteor shower A large amount of meteors originating from a common point in the sky.

meteorite A chunk of space rock that reaches Earth's (or another planet's) surface.

Milky Way The spiral galaxy that contains our Solar System. Its name is also used to refer to the faint band of light that can be seen on dark nights, composed of distant stars within our own galaxy.

moon A natural satellite that orbits a planet. When capitalized, used to refer to the body that orbits Earth.

multiple star A system of several stars that are bound together by gravity and all orbit the same centre.

nebula A cloud of gas and dust in space, visible either because it is illuminated by nearby or embedded stars, or because it is obscuring more distant stars.

northern pole star The alternative name for Polaris or the North Star, the star closest to Earth's North celestial pole, which is often used for navigation. It can be found in the constellation Ursa Minor.

observatory A place used for observing astronomical events. This includes ground-based observatories with large domes or dishes, as well as space-based telescopes and airborne observatories.

open star cluster A loose star cluster formed when a group of stars is born inside a nebula.

orbit The path a celestial object takes around another object when affected by its gravity.

phase The fraction of the Moon that is illuminated by the Sun, as seen from Earth.

planet A roughly spherical object that orbits a star and has a large enough mass to have cleared its orbit of debris.

planetary nebula A type of nebula formed from the gas shell cast off by a dying star.

Polaris (see northern pole star)

pole The most northern or southern point on the axis of a sphere. Earth's north pole is used for navigation.

pulsar The collapsed core of a large star that emits radio waves and other radiation as it spins.

red giant A large, reddish star with a low surface temperature that is reaching the end of its life.

red supergiant A large, bright star with a very low surface temperature. Red supergiants are the largest known stars.

shooting star (see meteor)

Solar System Our Sun and all the planets, dwarf planets, moons, asteroids, meteoroids, comets, dust, and gas that orbit it, along with Earth.

star A large sphere of gas that produces energy by nuclear reactions in its core.

starburst galaxy A galaxy where stars are forming much more rapidly than usual, often due to a collision with another galaxy.

star cluster A gravitationally bound group of between a few tens and approximately 1 million stars, all of which are thought to have formed from the same original massive cloud of gas and dust.

stellar Relating to stars.

Sun The star nearest to Earth that all the planets in the Solar System orbit.

supernova A violent explosion of a star which causes its brightness to increase enormously.

supernova remnant The outer layers and debris from a star that have been ejected during a supernova explosion.

universe Everything that exists, including all matter, space, and time. The Universe is thought to have begun in a big bang about 13.8 billion years ago.

white dwarf A small, hot, dense star, consisting of the shrunken remains of a large star that has ejected its outer layers.

zodiac A band of the celestial sphere that lies either side of the ecliptic, through which the Sun, the Moon, and the planets travel.

ACKNOWLEDGMENTS

DK would like to thank the following people for their assistance with this book:
Carron Brown for the index; Vicky Richards for additional editing; and Simon Mumford and Jacqui Swan for additional design.

Picture credits
The publisher would like to thank the following for their kind permission to reproduce their photographs:
(Key: a-above; b-below/bottom; c-centre; f-far; l-left; r-right; t-top)

2-3 Alamy Stock Photo: Design Pics Inc / Kevin G Smith. **4 Dreamstime.com:** Serge Bogomyako / Sergebogomyako (r). **5 Alamy Stock Photo:** Nature Picture Library / Jose B. Ruiz (r). **6 Alamy Stock Photo:** B.A.E. Inc.. **8-9 Alamy Stock Photo:** Image Source / Pete Saloutos. **14-15 Getty Images:** Stan Honda / AFP. **18-19 Getty Images:** Ian Forsyth. **24-25 NASA:** ESA and the Hubble Heritage Team (STScI / AURA). Acknowledgment: J. Gallagher (University of Wisconsin), M. Mountain (STScI) and P. Puxley (NSF). **34-35 NASA:** ESA, S. Beckwith (STScI), and The Hubble Heritage Team (STScI / AURA). **44-45 Alamy Stock Photo:** Stocktrek Images, Inc. / Alan Dyer. **48-49 ESO:** H. Drass et al (http://creativecommons.org/licenses/by/3.0).

60-61 NASA: ESA, J. Hester and A. Loll (Arizona State University). **68-69 Science Photo Library:** Babak Tafreshi. **78-79 Philip Perkins:** M31 image copyright Philip Perkins www.astrocruise.com. **82-83 Science Photo Library:** NAOJ / Hubble Legacy Archive / Judy Schmidt / Robert Gendler. **90-91 Till Credner / AlltheSky.com. **96-97 NASA:** ESA, and C. Robert O'Dell (Vanderbilt University). **108-109 NASA:** ESA / Hubble. **114-115 Alamy Stock Photo:** Stocktrek Images, Inc. / Alan Dyer. **116-117 NASA:** NASA's Scientific Visualization Studio / Ernie Wright (USRA) (b). **117 NASA:** (tr). **118-119 Alamy Stock Photo:** John Sirlin. **122-123 Alamy Stock Photo:** NG Images

All other images © Dorling Kindersley
For further information see:
www.dkimages.com